THE LECHER ANTENNA

ADVENTURES AND RESEARCH

IN

GEOBIOLOGY AND BIO-ENERGY

Written by Anne-Marie Delmotte

Second edition May 2019

Disclaimer: The author of this book does not dispense medical advice or prescribe the use of any technique as a form of treatment for physical or medical problems without the advice of a physician, either directly or indirectly. The intent of the author is only to offer information of a general nature to help you in your quest for well-being. In the event you use any of the information in this book for yourself, which is your constitutional right, the author and publisher or their authorised representatives assume no responsibility for your actions.

Second edition May 2019

ISBN: 9789082802672

EAN: 978-90-82802-67-2

NUR: 402

9 789082 802672 >

Cover photo:

Consists of

The foldable CEREB Lecher antenna® of which the picture is kindly provided by Michel Lespagnard of the non-profit organisation CEREB in Belgium

and

the Flower of Life symbol:
the Flower of Life consists of a series of evenly spaced interlinking circles. As more circles are added, the pattern emerges. The Flower of Life can be found in ancient temples and carvings around the world. The oldest depiction of the flower of life known to man is at The Temple of Osiris in Abydos, Egypt. The symbol here is not etched or carved into the granite rock, it appears to be burned into the rock. Many other depictions can be found within ancient structures in many different countries including Israel, Turkey, Ireland, India, ... A rather interesting depiction of the flower of life is found in The Forbidden City, in Beijing, China where the Flower of Life is on a sphere underneath the paw of the Fu-Dog. The Fu-Dogs were a symbol of protection to the buildings and their inhabitants and also, they were considered to be the guardians of knowledge which is interesting when you consider the spherical Flower of Life symbol that is found under the dog's paw. The dogs are literally guarding the knowledge. The Lecher antenna has been a very helpful tool in lifting the veil to this lost ancient knowledge.

All photographs by Dame Anne-Marie Delmotte unless specified otherwise.

A catalogue record for this book is available from the Royal Library in Brussels Belgium.

Preface:

I wrote this book because I would like the reader to know what this small but valuable instrument, that is the Lecher antenna, is able to detect and do. As a scientist it has been a real eye opener for me and I hope that for anyone who reads this book this will be an eye opener too. I have been amazed by my findings time upon time all through my Lecher antenna adventures. I am also overwhelmed by the discoveries I made; a knowledge I feel I must share in order for it not to get lost over time.

I wish that whoever reads this book will feel as amazed as I as to what this wonderful instrument can detect and do and that they will want to start their own personal journey. As you will be able to conclude from the cases, the Lecher antenna is certainly a very helpful instrument to have to help friends, family and yourself.

I would like to thank the following persons for helping me make this book a reality:
Melody Urquhart for her warm hospitality and letting me measure the energies at Gillighan's World Knocknashee and her never ending support.

Morag Donald for teaching me Reiki and spurring me on to continue with my research and this project.

John Willmott for his warm hospitality and letting me measure the energies of his labyrinth and triple spiral at Carrowcrory Cottage.

The members of my Facebook group "Lecher antenna – dowsing – bioenergy – geobiology" for their interest in the Lecher antenna and their support in my quest to write this book.

Anne-Marie Delmotte

CONTENTS

CHAPTER 1: THE LECHER ANTENNA ADVENTURES

1.1. BELFAST AND A BLESSING IN DISGUISE NORTHERN IRELAND

I'm in a most beautiful four bed family room in the Premier Inn Titanic Quarter in Belfast, although I only ordered a single room. It must be a sign that I'm on the right path. At Premier Inn peace and quiet is guaranteed. Everything here is in purple, my favorite colour. Carpet, curtains and walls. I went for a lovely afternoon walk to the old Harland and Wolff offices that have been transformed into a hotel and their opening day is today. I had the pleasure of visiting their lovely exhibition. Pictures reveal Harland and Wolff's huge success and the high number of people that worked here. Even carpenters, seamstresses, painters (of art), ... ; 20.000 according to a very old local man I had the pleasure of meeting and talking to and whose granddad used to deliver the timber for the ships, the staircase of the Titanic included. It's all gone now, at the bottom of the sea. Harland and Wolff is too, "not at the bottom of the sea of course", but gone, bankrupted; history. Should have built cruisers the old man entrusted me.

I did quite some travelling just to get here. Brussels to Dublin, then off to my wonderful house in Ireland to collect my Lecher antenna.

My home is situate amidst many cairns and stone circles that are thousands of years old. This Lecher antenna has two holes in its base, two holes are needed if you want to work with the ACMOS method. The right hole is to clear the energy from a person, the left one is to send the energy. I was reading Dr.Szulc's book "The Art of Medicine" about his ACMOS training adventure and experience with the Lecher antenna, and I thought I too could write about my own adventures and include my geobiology and other findings so these would not get lost as notes left on loose pieces of paper.

My mind has been very pre-occupied lately. I prepared to take a career break for quite some time now to be able to have more time for my energy project and doing up my Irish home and planned to give notice last week which I did. The reaction of my superiors was actually very positive. Retirement age has gone up considerably in Belgium and they think an early break to be a good idea. Finally, I thought, peace of mind and full steam ahead.

Only, two hours later an email came in from my Union (I'm a Union rep) about a new proposition of law cancelling all career breaks for government officials. The news hit me like an iceberg. I knew this was on the horizon for quite some time now, but no measures for people already in the system. Back to work full time for all of us. No pity. The country is in need of money and government officials are easy targets. So there goes all my planning for my career break. So, yes, I have a lot on my mind. Now I'm looking into stopping to work altogether and moving to Ireland. Maybe that's my blessing in disguise. Make the jump and don't look back.

1.2. BELFAST AND INVERNESS SCOTLAND

I took an early start this morning in Belfast, although the airport is only five minutes away from the hotel. I'm travelling with a soft top bag and was worried that my Lecher antenna would get bent in the haul of the plane if it stayed in the bag. I asked at the check in of my bag whether I could take my antenna as hand luggage, but I was asked to take it up with the security people at the hand luggage check. It turned out to be fine for me to take it as hand luggage so that was my Lecher antenna safe and sound.

Waiting on the plane I met a lovely couple from the States that were on a quest for places of energy. They had just visited Carrowmore, a megalithic cemetery in my Irish homeplace Sligo that is at least 5000 years old. The couple was now on their way to visit Stonehenge, Avebury, Salisbury and Winchester. Hmm, a coincidence, me meeting these people?

The weather has been on my side since I left my home in Ireland. It's autumn and getting a bit chilly as we are quite high up north. Some brief showers, ones that get you drenched in minutes. Luckily, I seemed to be spared from those as it only rained when I was inside somewhere. When we got on the plane, a purple one by the way, there's my favorite colour again, a most beautiful rainbow starting from just right behind the plane appeared in the sky. A sign that I am surely on the right path.

Once in Inverness, a wonderful lady, Elizabeth picked me up. It's as if she was heaven sent. It was so difficult to find decent accommodation for a reasonable price in Inverness and her place just dropped out of the sky for me one day. She had just put up her place on AirBnB, wanting to give this a try because she wanted to meet new people. I just knew that this was the place where I needed to stay during my ACMOS training course.

I headed for Inverness in the afternoon for a brief visit. Luckily the sun was out and it was a very clear day. The views from the top of Inverness Castle were stunning. I lit a candle in the Inverness Cathedral and had a brisk walk along the river. I couldn't resist buying a new raincoat, a purple one of course, my favorite colour, remember. One for walking in the Highlands, well at least in Ireland, it has a lovely tartan print on the inside. A rare find. The local buses here are like the ones in Ireland.

Noisy and very old, but charming nevertheless. I'm back at my AirBnB now and Elizabeth is just so nice, she's letting me have access to her living room, dining room and kitchen. It's a lovely tidy place and guess what. She had a Lecher antenna treatment years ago and asked me if I would give her one.

Tomorrow is my first day of the ACMOS Lecher antenna training. Somehow, I feel that I'm on the verge of something big.... A turning point in my life...

1.3. BALLOCH ACMOS DAY 1 SCOTLAND

I got up very early today to do a self-Reiki session first in order to make sure that all my energies were optimal for my training course.

It was a dry cloudy day so I decided to walk over to where the training course was being held, our trainer, Carol Robertson's house. A small path took me next to the woods and then on to Culloden Avenue that has wood carvings of dragons, dragon-eggs, a faerie and other wooden curiosa. On the way a man called out "Cinderella", not sure whether he was talking to his dog or his daughter. What an odd name to be blessed with. Once at Carol's house I was greeted by her barking dog, Lola, who turned out to be actually very nice and was just looking for attention in order to play with her.

As this is level one of the training course a lot of what we looked at today was very familiar to me. We learned what the Lecher antenna settings on the ruler of the antenna 8; 12; 15.3 (see glossary page 95); 17.6; 2.1; 2.5 and 1.1 mean in bio-energy readings. We then measured the quality of the room and each other. We also tested objects that we were wearing and corrected them if needed. We did a compatibility testing of these objects and ended with an introduction of the meridians and acupoints.

I pointed out to Carol that my watch is disturbing my energy-fields and that I correct this by wearing a shungite crystal. She suggested to do a full ACMOS later in the week to see whether this can be corrected otherwise.

Something very important that I withheld from today's course is that you should never go into the negative by telling people that they have an entity or negative energy on them because this will only trigger fear into people and feed wrong energetic programming.

After the course Carol and I took her dog Lola for a long walk in the woods and fields and we had a lovely chat about our own life's challenges. She even drove me back to my accommodation after our walk. A lovely way to end the day.

I still need to do some preparing for tomorrow. Carol gave me a purple folder in which to file all our documents which we will need for the full ACMOS protocol.

1.4. BALLOCH ACMOS DAY 2 SCOTLAND

It was raining this morning so I decided to take the bus to the training course in Balloch. The windows of the bus were so damp that you didn't have a clue as to where you were and when to get off. I used the sleeve of "me bonnie new purple raincoat" to wipe the window and managed to catch a glimpse of a street sign that says Culloden Road where I needed to get off.

Carol's dog Lola greeted me with her usual barking to get attention.

The course started with measuring each other's Ting points (special acupoints) on our hands with the Biofeedback tester. The purpose is to check these points again after "balancing" the person for improvements. We were then taught extensively about the body's energy fields; the magnetic field at 5.7 on the Lecher antenna and the electric field at 7.8 on the Lecher antenna and how they can be measured together at 15.3 on the forehead. After lunch we learned how to use the Acmograph. This graph is used to measure how much defense energy is available, the values of the electric and the magnetic field, the mental state, the different balances (organic, lymphatic, vibrational, mental and hormonal), the quality of the habitat and how a person is in his or her habitat.

Carol then measured me, although there was hardly anything to balance. My 15.3 on the forehead was off by 3 centimeters to the left which is almost insignificant. The main reason I wanted to be balanced was to strengthen my energies so that I would be able to withstand the disturbance of my phones and of my watch without having to carry a shungite crystal in my pocket to correct. With my watch in the circuit my 15.3 measured on my forehead was 12 meters off to the right. Carol said that she sensed the disturbance as well on her chest (heart). After going through the protocol the best thing for me to do was to get rid of the watch altogether because in that configuration the quality of my fields were the highest and way off the chart. I didn't do any measurements with the antenna today which was a bit disappointing. Almost none of the students studied anything beforehand which is really slowing us down. A bit of a shame because we lose good quality time due to this. We cover a large amount of information in a very short time and on top of that the students that have never handled an antenna really need to get the hang of it. We received the course documents 6 weeks in advance. It does contain information about how to use the antenna and how to do simple exercises.

We then went over how to use the Acmograms and the other students worked on acupoints and opening the drivers 8 and 12 (see glossary page 95) which can only be opened if permitted by the energetic body of the person. I was amazed at how many blockages the young students had in our group. I seemed to be passing most readings with flying colours. I think this might be because of my sugar free and carbohydrate low diet, my self-Reiki treatments and the energy balancings I'm doing for this book as these seem to have a beneficial effect on me as well.

Once back in my AirBnB my host Elizabeth welcomed with a poem she wrote about me and I just gave her an energy reading in return. At least this way I put my antenna to some good use today. Here's the poem the very Scottish Elizabeth wrote for me today:

THE RAINCOAT

by Elizabeth Norris

A lady came from Belgium to the city of Inverness
She was very pretty and fashionable in dress
She stayed here at Benarty, went on a shopping spree
Looking for a raincoat, necessary I agree.
A purple raincoat she did buy with inner tartan layer
She looked so very smart in it an answer to her prayer
Scottish weather can be wild and soak you to the skin
You really need a raincoat from shoulder to the shin
When you come to Inverness always dress for rain
With beautiful green scenery you're welcome back again

1.5. BALLOCH ACMOS DAY 3 SCOTLAND

My day started with doing some more Lecher antenna readings. This time I measured the bed of my host at my AirBnB place. I found a crossing of two Hartmann and one Curry line on the side of the bed where Elizabeth, my host has been sleeping every night for the last ten years; this crossing is at the height of her heart region. She needed a major heart operation a few years ago and this crossing of energy lines is certainly not beneficial to her. She seemed very keen on wanting to continue sleeping on that side of the bed and did not really have the possibility to move the bed either, so the idea came up to use crystals. I had recently read that amethyst and rose quartz are good for correcting. Elizabeth likes crystals very much and even followed a training course about them so I asked her whether she had any. She did indeed have an amethyst and a rose quartz among many other crystals. I put the rose quartz and the amethyst on the spot where the energy lines cross and the two Hartmann lines and Curry line now no longer gave a signal when I scanned with my antenna. It looked as if the lines were gone, well at least dissipated. I re-checked after putting the crystals under the bed and this was good too. I then checked how she needs to regenerate the crystals and the answer was to leave them for one day in the sun every three months. A success. I wanted to try the exact same thing for a few months now so I was happy that this opportunity had presented itself to me and that I had a positive outcome.

It was a dry morning so I decided to walk to my training course again. On my way I was treated by more lovely wood carvings of animals, mushrooms, butterflies and dragonflies which I hadn't noticed on my first walk out here two days ago. I also witnessed a most beautiful rainbow. Another sign that I was on the right path?

Today we went over the marvellous vessels and the many other acupoints that are used in the ACMOS method. I then was given the opportunity to balance one of my co-students; the person's 1.1 was closed which is the connection between the deep centripetal Yin field and the outward centrifugal Yang field or in other words, the gateway or bridge to the energy reserves. This is a very important frequency and may NEVER be opened by imposing. I went through the entire ACMOprotocol which lead me to stimulating 2 pairs of acupoints and balancing an ecological problem, this being a physical or mental relationship issue. In order to work on the latter the antenna told me to reinforce myself first. I needed to put one tablet of Acmochel, a Chelidonium homeopathic, on my body. Hmm, a coincidence. Over the last few weeks I had been measuring many of the harmonies needed for an ACMOS balancing with my Lecher antenna which were kindly provided by my physiotherapist and the Chelidonium was one of the very few products I didn't have. After reinforcing myself the antenna lead me to a colour I needed to use to balance my co-student. After the colour balancing of the person no further interventions were necessary. The bridge to the energy reserves (1.1) of the person had now opened and there was an improvement of the person's mental state.

We then went over the environment protocol and discussed some of the more rare products. We got to see Carol doing an environmental protocol for an energy leak on one of the students. This took the rest of the afternoon and we got to witness Carol's famous way of acmodynamising. I asked what her secret was into moving the antenna so quickly in order to clear or put in energy and she told us that she was able to make the antenna move on its own after years of practice. She also taught us how and at what setting we can strengthen or clear negative memories after an acmodynamisation.

We had a very long training day and I felt I had learned quite a lot of new things that I might apply in my balancings for my research.

Tomorrow is my last training day. I am now at my AirBnB writing this and going over my notes. It has been an intense number of days up until now.

1.6. BALLOCH ACMOS DAY 4 AND VISIT OF CLAVA CAIRNS SCOTLAND

It was a dry morning so I decided to walk to Carol's house again. I noticed the birds chirping lively just before I got there. I was too early so I took a further walk in the woods and field just behind Carol's place where I found a four-leaf clover. I asked the universe who in our group I should give the four-leaf clover to and Carol's name came up. I put my Reiki intention of love, light and luck into the clover and gave it to Carol who was very happy with her gift. Some of the students said that they had never seen a four-leaf clover and took a picture of it. I seem to find them all the time. More signs that I'm on the right path?

Our training course started with a Powerpoint presentation which was an overview of the ACMOprotocol, what all the boxes mean, and how to proceed when the Lecher antenna gives a signal on one of these boxes.

Interesting to know is that if a product has lost its 1.1 (see 1.5.) it might come back on if you impose the 17.6 on it. The 17.6 has a regenerating quality. Carol says that the products keep all their frequencies for a very long time and that they harmonise each other when being put together.

Carol then demonstrated a balancing on one of the students. Some of the time she was just going over the charts in her head and scanned them as if they were projected in the middle of the room. Some of us worked with the Acmodermil, a device that is used to stimulate or sedate acupoints. The output signal is very strong so it actually works on the entire meridian.

In the afternoon we worked on objects of the other students that had no 2.5 (minimum requirement for object to have) and checked their compatibility, in order to put the object back to the best energy for this person. We then balanced an energy leak on each other. The student's energetic body I balanced asked for an acmodynamisation of an essential oil, cedar to be exact. After the acmodynamisation I strengthened an elimination memory and further cleared a negative memory which is inherited, or genetic, that dated back to May 1760 in the Balloch-Culloden region. This would not be long after the battle of Culloden at Culloden Battlefield of which the memorial I hope to visit tomorrow. Carol told us that these were very harsh times for the Scots. I have to admit that I felt shivers down my spine three times when clearing this. I went on to clear a negative emotional memory and finished the balancing of the leak with a colour. After this I checked the student for improvements after the balance. The student I treated then treated me for an energy leak. The oil that came up for acmodynamisation for me was mint, one of the few oils that are related to the hereditary and anteriority reducer, the most important reducer of a person's energy. After the acmodynamisation a negative memory was cleared from my cells and also a close physical relationship issue dating back to January 1994. The only thing I remember from that time is that I gave "notice" for a better job. There were major changes happening at the company I worked. There was a change in management and a lot of people I had a good relationship with were laid off. I remember that I had difficulty coping so the new job was heaven sent. It was indeed a turning point in my life. After the balance my mental state improved from cheerfulness to serenity.

I thanked Carol for the interesting training course and said goodbye to my co-students hoping we would all stay in touch through Facebook.

One of the students was kind enough to take me to see the Clava Cairns at Balnuaran of Clava after the course where we did some geobiology readings with the Lecher antenna.

Clava Cairns is a well-preserved Bronze Age cemetery. It consists of three large burial cairns, each surrounded by a stone circle. Two of the cairns are passage cairns, where the central burial chamber is accessed via a narrow corridor. Between the two passage cairns sits a kerb cairn, with an enclosed central burial chamber.

I found a Leyline at 16.4 coming from the entrance of the passage cairn which is the most North one of the three cairns. It was the same kind of Leyline as the ones I had discovered in Ireland between most of the megalithic structures (see later in this chapter). This 16.4 Leyline I first discovered at Maeve's Cairn on Knocknarea in Sligo (see 1.15.). I also found sacred energy (= 1.1 on the Lecher antenna) at the entrance of the same cairn in Scotland; in other words sacred space.

I managed to check the Hartmann and Curry grids (see chapter 5). All the Hartmann and Curry lines have moved and are heaped up just outside the stone circle surrounding each of the 3 cairns like at the pyramid of Gizeh and other sacred sites (Blanche Merz; see also 1.8.) and like in my quartz-containing stones experiments (see chapter 2). I also found a 16.4 Leyline between the 2 big standing stones that are on both sides of the kerb cairn (outside of its stone circle).

Picture: split stone at Clava Cairns

Picture: kerb cairn at Clava Cairns

We met some tourists from Australia who asked what we were doing and I explained that we were measuring the energy lines. One of the men in the group said that he felt "electrical". I asked if he was okay with me doing a little check which he was. His electrical field was off to the left. I asked whether he was wearing a phone or a watch. He had a watch that works on a battery. His electrical field was centred without the watch. I asked whether he wanted me to energise the watch so it would be compatible for him to which he agreed. I energised it at 2.5 which was the value his energetic body asked. I imposed the 2.5 on the watch and the man's electrical field was now centred with the watch in his hand. Another good deed done. I do seem to have a thing with watches. I'm now at my AirBnB writing this and drinking peppermint tea after having a lovely chat with my host Elizabeth.

Picture: doing Lecher antenna readings at Clava Cairns Scotland

1.7. CULLODEN BATTLEFIELD - INVERNESS SCOTLAND

I set out this morning to visit Culloden Battlefield, but for some reason fate decided otherwise and my bus didn't come. It was probably better for me anyway not to visit this place.

The battle at Culloden was the last great battle fought in Scotland for who would be on the throne in London. An estimated 1500 Jacobites died on this battlefield - men who had worn the blue bonnet and white cockade with pride. And trampled into the boggy, blood-stained ground that awful 16th April 1746 day, alongside the discarded broadswords and targes of the clansmen, was the last, lingering hope that the ancient Stuart dynasty might come to reign again.

Headstones were put on this battlefield in memory of the clans and clan members that lost their lives that day. I assume many of their bones still remain there.

I remember I visited the trenches in Diksmuide in Belgium from the first World War many moons ago on a very hot summer's day and that I wasn't able to walk through them for very long as I was freezing from the chills I felt running down my spine. It was the same feeling I got yesterday when I was clearing away the inherited/genetic memory from 1760 during the balancing of one the other students.

After the battle at Culloden, the terror continued. Parliament decreed that Jacobite chiefs were stripped of their estates and their powers of jurisdiction. Their clansmen were simply stripped, forbidden by a petty Disarming Act from carrying weapons and wearing kilts – even playing the bagpipes and speaking Gaelic!

It looks like there are still a lot of energy balancings to be done to clear people of these negative inherited imprints.

I think that bus did not come for a very good reason. I was not meant to visit Culloden Battlefield. It just might be too much for me energetically.

I set out to Inverness town instead where I bought a book explaining the Culloden Battle in detail. I asked the man at the bookshop about how accurate the facts of this battle are in the popular TV series Outlander. He confirmed that the original books were not correct, but that a historian and a Gaelic specialist were brought in to make sure these facts were now as they happened in Scottish history for the filmed version of these original books.

I also bought Elizabeth a little ladybird which I energised with my Reiki intention of love, light and luck as a thank you for being such a wonderful host. And guess what, I found myself a nice new purple digital watch. I really wanted closure on it whilst I was still in Scotland. When I saw this shop I just knew that they had the watch I was looking for. The watch has 2.5, but I brought it up to 8 because that was what my energetic body wants the watch to be. The 8.0 has a calming effect on the person. This watch has no influence of my magnetic and electrical field at all like my old watch that Carol tested on me earlier this week. Even without the crystals I'm carrying.

I rushed back to my AirBnB to collect my bags. Elizabeth gave me a lift to send me on my way. She said she got really used to having me around and she couldn't believe I was already leaving.

I am now at Inverness airport waiting to be flown back to Belfast on another purple plane.

1.8. CARROWMORE MEGALITHIC COMPLEX SLIGO IRELAND

I woke up this morning in my favorite home place, Sligo, and noticed that there wasn't a cloud in the sky. I got up very early to watch the sunrise in the field just beside my house that has a view of all the mountains around us. Most of these mountains have cairns on them. It was very quiet and the only sound I could hear was of the birds chirping away very cheerfully in the distance. The days are longer here than in Scotland. Sligo is less North than Inverness. I love this time of year in Ireland.

After cutting the grass I set out to do some more readings of local stone circles with my Lecher antenna. I discovered two readings at Clava Cairns in Scotland and wanted to check whether these could also be measured in Sligo.

I am so fortunate to be able to live in Sligo. There are still so many Megalithic Monuments. Cairns, court tombs, dolmen, stone circles, passage tombs, … . Even my address has the word cairn in it, twice. Well, my house is located amidst major cairns.

It was an Indian summer's day. The sun already felt quite warm, even if it was still early morning. I decided that my best chance for stone circles would be Carrowmore megalithic cemetery.

The Carrowmore megalithic cemetery covers an area of about one square kilometer. Most of the tombs have been arranged in an oval-shaped lay-out, forming a well-defined ritual landscape. There is a tendency that the tomb entrances seem to face the central part of the cemetery. None of the tombs at Carrowmore display the kind of passage that normally features in Irish passage tombs, and only one (tomb 51; Listoghil) shows remains of a cairn.

I measured the major tomb called Listoghil first, number 51. I wanted to find out whether the Hartmann and Curry grid lines (see chapter 5) move outside the cairn, clearing the inner space of their energy. It seems the Hartmann and Curry lines are all heaped up just outside of the structure here too. Blanche Merz (Swiss building Engineer and Geobiologist) states in her book: Blanche Merz has found the Hartmann grid to be pushed outside certain sacred structures such as the Egyptian pyramids and temples and Himalayan Buddhist stupas, creating a dense protective wall around the perimeter, and a ray-free interior.

Picture: tomb 51; Listoghil

The Leyline at 16.4 on the Lecher antenna at the entrance of the cairn that comes from the stunning cairn topped mountain of Knocknarea that is 6 km to the west, is not deviated and does run within the Listoghil tomb and then goes on to the Ballygawley mountains. I checked for sacred energy at 1.1 on the Lecher antenna as I found this energy by coincidence at the Clava Cairns in Scotland and found a signal. After investigating further I discovered a grid of 1.1 lines. 1.1 stands for sacred energy/space. The direction of the 1.1 was towards the inner space of the cairn in which a central chamber stands.

Picture: Tomb 51: central chamber within the Listoghil cairn

Earlier measurements with the Lecher antenna indicate that the central chamber is standing on a major water source. I got 39 signals for water (at 7.8) around the outer part of the cairn and 15 signals for water (at 7.8) around the central chamber within the cairn. I remember finding a similar reading of several waterlines in the Cathedral of Chartres in France near Paris.

After this I measured 4 different stone circles: the 57, 1, 3, and 27. They all showed the Hartmann and Curry grid free space within the circle and these grid lines are heaped up just outside of the stone circles. Circle 1 showed a spiral signal (see glossary page 95) of a huge Curry line deviated outside of the circle. Inside these stone circles I did not find any 1.1 sacred energy.

Picture: stone circle 1

Picture: stone circle 3

1.9. BACK TO KNOCKNASHEE AND GILLIGHAN'S WORLD SLIGO IRELAND

Today I will pay a good friend of mine a visit. Her name is Melody and she is the owner of Gillighan's World, The Field of Dreams, which is located at the foot of Knocknashee, the Gaelic for "Hill of the Faeries". Melody is a wonderful and very courageous person. She came over from England many years ago to set up Gillighan's World in order to bring an almost forgotten ancient site, its legends and folklore into the modern world. Gillighan is the name of the 19/20th century owner, and has nothing to do with Gillighan's Island, a popular TV series. A visit to Gillighan's World is a spiritual experience for people of all ages. It is an escape into a land of artistry and imagination, and with its indisputable "energies" and "spiritual healing" ambience it is a place adults adore, whilst children love the magical setting for the faeries. Miniature villages and delightful faerie habitats nestle in beautiful water and botanical gardens. A stone-tunnel entrance, dolmen and amphitheatre give the park a sense of antiquity befitting an ancient site, augmented by an enchanted glade and cave, secluded picnic areas, a pet haven, refreshment area and shop. There are spectacular views from the Celtic Faerie Fort that Melody restored.

Knocknashee is a huge fortified hilltop, although it was always known as a place of great interest because of the two cairns. The site is situate on a table top plateau with stunning views of the Connacht plains and Ox mountains. I remember doing Lecher antenna readings high up on Knocknashee in 2016 and discovered that there are two 16.4 Leylines coming from one of the cairns, one of which is coming out of the entrance of the cairn. I took compass readings for their direction and it would seem that the line that is towards the North (that is 23 degrees wide) runs all the way to Maeve's Cairn. There is also a 16.4 Leyline towards the South that is 5 degrees wide coming from the cairn.

The Leyline that is running North also encompasses a large area of the Gillighan's World which is down below on the mountainside and also runs through the restored Faerie Fort (left side of entrance of the fort) at Gillighan's World.

Picture: Knocknashee and the Faerie Fort at Gillighan's World in the front

I found several 16.4 Leyline signals at Gillighan's World which could be the spiral signal of the wide line coming from the cairn's entrance on the Knocknashee table top. A similar very wide 16.4 Leyline is also coming through Gillighan's World from Maeve's Cairn on Knocknarea. In a particular spot at Gillighan's World the Leylines from Maeve's Cairn on Knocknarea and the cairn on Knocknashee meet because the Lecher antenna is indicating both directions as the origin of the energy. I have found the very same type of reading between the Listoghil cairn at Carrowmore and Maeve's Cairn on Knocknarea when measuring at Carrowmore between both cairns. This could be an indication that these straight lines seem to form between the cairns (two structures of the same type and building materials) and that there is a back and forth energy interaction between the cairns.

Last year Melody moderated a weekend visit for a group of professional paranormal investigators interested in the energies and vibrations emanating from Knocknashee, and daring to hope to catch a glimpse of the faeries. She is very "open" about her experiences at Gillighan's World and admits she has often seen two "beings" up on the mountain particularly at twilight, and she is not alone in regard to these strange sightings. A picture of one of these beings was taken by local author and renowned paranormal investigator, Barry Fitzgerald, who has done a lot of research and work in the subject of the Sidhe, the Gaelic word for faeries. Melody also told me of what happened on the second evening during this particular weekend which the small party of invited guests got to witness. In fact the activity exceeded their wildest dreams, dancing lights and huge orbs were sighted, but Melody and another couple caught a glimpse of another world, with huge buildings, gigantic trees and doorways (portals) and the spectacle lasted for almost two hours. Measurements with my Lecher antenna reveal an interdimensional portal at 14.7 which is the setting for interdimensional energy and several 16.4 Leylines which looked like this at the location where the "other world" was witnessed.

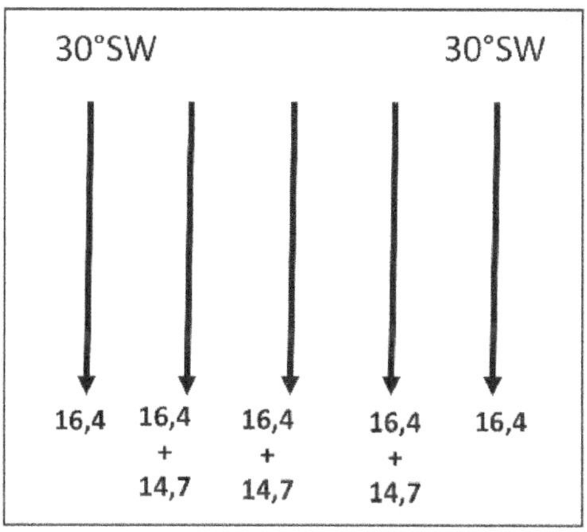

SW=Southwest

I was back out there several times and these interdimensional energy measurements look stable, at least measurable each time I did a reading. The lines can be followed further down a field where it is on the land of another owner. Melody told me she will ask for permission for me to go on that land to be able to trace these lines further down the fields and see where it leads me....

HILL OF FAERIE

by Anne-Marie Delmotte

Oh Knocknashee
Hill of Faerie
The secret held by thee
You have shown me

An interdimensional door
To another realm
Who could ask for more
You overwhelm
All who come to Knocknashee
To look for Faerie
And feel your energy

Oh Hill of Faerie
Knocknashee
I thank thee
For sharing your secret with me

1.10. CREEVYKEEL COURT TOMB CLIFFONY AND MULLAGHMORE IRELAND

End of september and yet another Indian summer's day in Sligo. I used the opportunity to go back to a court tomb in Cliffoney where I did some measurements in June 2016. I wasn't aware of the Leyline at 16.4 at that time and I wanted to see whether I could find the same energy here too. The court tomb is called Creevykeel. It is one of the finest Court Cairns in Ireland, built in the 3rd millennium BC. It consists of a wedge-shaped mound, with a burial chamber opening off the open central court. Near the narrow end of the mound are two further graves (passage graves?). In the early Christian period iron was smelted in the central court. Some of the stones still show signs of iron.

Picture: Creevykeel court tomb seen from its entrance

I found myself alone at the sight and although it was a sunny day it felt very chilly. I went to fetch a shawl and my anorak from my car. When I took out my antenna it suddenly got very warm which made me take off the layer of clothing I had just put on.

A bus stopped at the site and I now found myself with about forty people from Canada on a 2-week tour of Ireland. They did not stay very long and soon I was alone again, only to be joined shortly after by a couple from California who loved visiting old sites. I got some tips on some very difficult to find tombs up in Donegal that are not kept, nor signposted, but with special energies nevertheless.

New people came to the site and it suddenly went very chilly again. This is not the first time this has happened. Just wondering whether this has to do with being welcome or not. Strange how this chilly feeling always goes away when I get out my Lecher antenna.

I finally was able to do my measurements. Here too, the Hartmann and Curry grid shift to just outside the stones where the lines are all heaped up. See chapter 4 for more information about the grids.

There is a 16.4 Leyline running through the central part of the court tomb, there is also a lesser wide energy line of cosmic energy (12) that runs in the middle of the 16.4 Leyline. The lines come out of the entrance of the tomb.

All the open spaces within the court-tomb did indeed show the 1.1 (sacred space) grid as within the Listoghil Cairn at Carrowmore. I finally had enough space to find out more about this 1.1 sacred energy grid. It looks like it is a 50 degrees Northeast grid (and Southwest) with lines that are about 2 meters apart and with a width of about 20 cm. Unfortunately, I did not think of bringing a measuring tape with me.

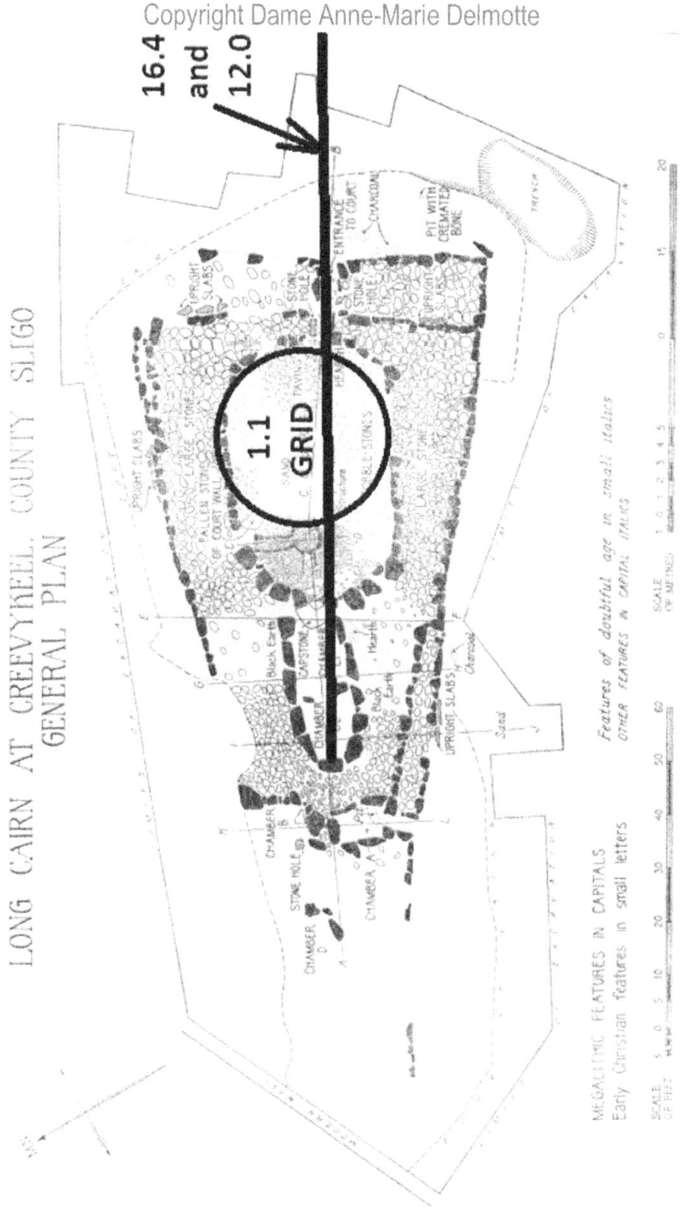

Picture: ground plan of the Creevykeel Cairn

After spending more than 2 hours at the site I was happy enough to be on my way and to further enjoy this lovely sunny day.

I went to Mullaghmore, a lovely little village by the harbour that has many palm trees and gives you the impression that it's summer all year round. I even saw a caterpillar. A bit late for this time of year? Not in Mullaghmore. I had a lovely walk around Mullaghmore Head enjoying the very relaxing sound of the crushing waves of the Atlantic Ocean. Surfers, the best in the world, come here to practice because of the infamous waves that can be up to 35 meters high.

1.11. LABYRINTH AND TRIPLE SPIRAL CARROWCRORY IRELAND

Today is the first of October. Autumn in Ireland, although we still can have lovely sunny days. I'm in luck because it's supposed to stay dry today with heavy winds picking up in the evening. A good friend, John Willmott, is having an open labyrinth day today at his house, Carrowcrory Cottage near the caves of Kesh.

He calls these days "Find your Story" days because he lets people go through a special ritual before and when walking the labyrinth. Nothing airy faerie, we all just get an apple that we cover with fresh ash, which we walk through the first part of the labyrinth with. We then wash the apple before we start the second and last part of the labyrinth.

After walking the labyrinth we all go sit around the fire and by means of a talking stick we each tell our story that came to us during the labyrinth walk. My story is to always trust my instincts.

Picture: talking stick and labyrinth apples

I set out very early to make sure I had the labyrinth and triple spiral all to myself to do some Lecher antenna readings. I had no clue what to expect, but I do know that labyrinth and spiral like patches are used to reduce the electromagnetic radiation from your phone.

This is what John asked me to include in this book about his labyrinth and triple spiral:

The Tree Labyrinth is a labyrinth of trees to encourage connection to trees and how they relate to the flowers, fungi, birds and insects around. It is therefore about tree and nature connection whilst also walking the spirals of the labyrinth.

The Triple Spiral Labyrinth is a more open and wild labyrinth, with views and semi moorland fauna. The main plants are kept simple; lavender, lemon balm, heather, rosemary, clover, and on the outside blueberries, raspberries and strawberries.

The Tree Labyrinth is more for deep contemplation, mindfulness and connection and the Triple Spiral is for fun and playfulness.

Picture: triple spiral

I walked inside the labyrinth and the triple spiral and scanned for Hartmann and Curry lines and found none (see chapter 5 for more information about the grids). I looked for sacred energy (1.1) and divine energy (17.7) but also negative. I checked for a vortex with my pendulum but also negative. I then asked the antenna and the pendulum whether there are any energies present in the spiral and labyrinth and the answer was no. As it seems, the labyrinth and the triple spiral are energy free zones.

All the Hartmann and Curry lines are heaped up just outside the last circle of stones creating a wall of energy I was able to see and photograph with my Harry Oldfield filter (see picture).

Picture: labyrinth taken with the Harry Oldfield filter; notice the wall of energy at the edge of the labyrinth

People were starting to arrive so I ceased my activities, although they were very interested in my findings and the Lecher antenna. I met some wonderful, spiritual people today and I am the better for it. We also enjoyed some great organic food which John had made for us and that the others had brought along. Some even brought freshly made apple crumble and whipped cream. One apple crumble was still warm, makes you wonder how fast some people drove to rush out here.

1.12. BACK TO CARROWMORE - SAMHAIN ALIGNMENT SLIGO IRELAND

I was awake very early this morning and saw that the sky was bright and full of stars. I decided to take my chances to go and see whether I could catch the sun shining in the dolmen in the Listoghil cairn at

Carrowmore in Sligo at sunrise. This only happens around Samhain (31 October).

It was a beautiful but a somewhat chilly morning. The valley was filled with a magical mist. There were some clouds about but most of the horizon just above the Ballygawley mountains was clear. The land had its familiar smell of moss. My house smells of this too. To me it's the smell of Ireland, of home. I did not see the first sunrise from behind the mountains, but the sun did show herself shortly after. I witnessed the sun shining in the dolmen firing up the underside of the roof-slab and lighting up part of the backstone. A very special experience. I've been coming here for days on end for years to see this and this is the second time I got lucky. So beautiful, so peaceful. The land was looking glorious. You can see all the cairns on all the mountains from here.

I found this to be the ideal moment to check a theory I recently found on dolmens being orgone generators:

"Another fascinating thing to know about many of the dolmens in Ireland is that originally they were entirely covered by alternating layers of organic and inorganic materials. While the purpose of this construction technique is presently unknown, it is interesting to note that the scientist Wilhelm Reich used the same technique in the construction of his so-called orgone generators, these being (much smaller) devices that were able to generate, concentrate and radiate a mysterious form of energy. Could the ancient builders of the dolmens have been using their unique construction techniques for a similar purpose? Orthodox archaeologists commonly assume these dolmen structures were used for funerary purposes because burials have been found in a small number of them (a very small number!). It is important to note, however, that the scientific dating of the burial remains show them to be hundreds or thousands of years more recent that the structures themselves, thus casting serious doubt on the tomb theory."

See also 1.14 about orgone.

I already mentioned earlier in the book that I found a sacred 1.1 grid within the confinement of the outer construction of the Listoghil cairn. I now checked for prana at 5.7; chi at 11.68 and transformation at 18.00 and found them all to be present. I could hardly believe it, so I checked whether these signals were genuine with the magnetised rod in negative in the left arm of my Lecher antenna (see glossary page 95). Yes, genuine; prana, chi and transformation energy and the 1.1 sacred energy. All these signals are coming from the opening of the dolmen (see picture). This is also the part where the sun shines through to the backstone and roof-slab at Samhain.

Picture: dolmen in Listoghil cairn seen from entrance

1.13. HOLY WELL SLIGO IRELAND

The day was still young and the sun was out and slowly warming the land. I could see the mists slowly dissipate in the distance. What better time than the present to look for the energies at the Holy Well in Sligo. Maybe I could discover another sacred grid at 1.1?

Once there I started walking around and looking for a 1.1 (sacred energy) signal. I got a signal at the height of a very ancient stone that was there before anything else was put there like the altar, statues of Jesus, Mary, the stations of the cross, ... A couple arrived and they each put their hand on the stone before heading towards the altar. I asked them why they put their hand on the stone. The man answered that there were finger-holes or cup marks on the stone and when you put your fingers into them it brings you blessings. I further investigated the stone and found there to be a crossing of two 1.1 lines (see picture), each of them heading towards the stone. One 1.1 line could be traced to a wishing holly tree that had all kinds of objects hanging from it which had been left behind by the people. Things like ribbons, crosses, glasses, hairpins, ... The other 1.1 line I could trace back to the statue of Mary that is high up on the right side of the stone and altar (see picture). After doing further measurements I found the Curry and Hartmann grids to be normal. I also discovered a broad, about 80 centimeters, divine energy (17.7) band going to the ancient stone and coming from the height of the same Mary statue. The only other time I found 17.7 energy was at the Cathedral in Chartres.

These are quite some findings for just one day. I feel amazed, overwhelmed and blown away. Certainly food for thought and further research.

Picture: cup marks on ancient stone

Picture: Mary statue from or behind where the sacred and divine energy is originating

Picture: the divine and sacred energy going to the ancient stone; the altar at the site is to the right side of the picture; the fingermarks in the stone is where the top of the Lecher antenna is (same position as the above picture of the cup marks)

1.14. CARROWKEEL MEGALITHIC COMPLEX BRICKLIEVE MOUNTAINS IRELAND

It is 31 october today. Samhain in Ireland, later to be known as Halloween. It is said that the veil between the spirit world and the living world is at its thinnest today.

It is a mild and dry day and I decided that it is high time to go back to Carrowkeel in the Bricklieve mountains near Castlebaldwin.

The impressive passage tomb cemetery of Carrowkeel is located above the western shore of Lough Arrow and located in the beautiful limestone uplands of the Bricklieve mountains. There are 14 cairns located at different prominent positions on the hilltops here with a further group of 6 cairns extending to the west to Keshcorran

Mountain, which is also capped with a large cairn. From here you have magnificent views of the surrounding countryside, principally to the north and west and many of the hill and mountaintops visible from here are capped with cairns. Dating has indicated the cairns here were built between 3800-3300 BC.

It was very windy going up the path leading to the cairns. I saw some lovely hawthorn trees in their autumn beauty of red berries and covered in a light-green moss that looks like angel hair on its branches. The moss is apparently a sign that the air is very pure. I have to admit that I notice the air here to be even purer than anywhere else in Ireland, although the air on the West coast of Ireland is already much purer than on the continent.

I also regularly captured a very particular smell (almost like a kind of natural antiseptic) on my way up to the cairns that I could narrow down to the withering bloom of the heather.

Once up there I had a lovely chat with an American person that I had given directions to earlier. As it seems the party of three were travelling all through Ireland trying to unravel the mystery of the megaliths. I told them about David Cowan's theory about dragon lines; about how they were placed on particular spots to neutralise the negative energy. I told them about my research and that I was here today to measure whether energy is being transformed here into positive and beneficial energy called orgone.

I am planning a visit to the Loughcrew cairns here in Ireland next spring and found out that orgone had been detected at Loughcrew (see also 1.12.).

I bought three orgone generators to familiarise myself with the orgone energy and was kindly provided Lecher antenna settings to research by a lady from Canada called P. who gives Lecher antenna training courses. She told me to look for: prana 5.70; chi 11.68; golden ratio 1.618 and transformation 18.00. I found the orgone generators to be positive for all these Lecher antenna settings.

I now checked these same Lecher antenna settings at the entrance of two of the cairns at Carrowkeel and found a positive signal for all of them which I double checked with my magnetised rod (see glossary page 95). Yes, indeed, it seems that orgone is being produced by these cairns, well at least chi and prana are!

At one of the cairns I met Jackie who lives in Ireland. She told me that she comes here regularly to take in the energy, particularly the strong energy felt within the cairns. I had felt reluctant to go into one of them until now and Jacky said she would assist me in doing so. We went into cairn G where I found ten kind of pillars (orthostats), of which eight form a circle and two are at the entrance. I detected a genuine signal for prana, chi and the golden ratio within the cairn. An indication of life force energy or natural energy. Transformation energy 18.00 was coming from all of the ten orthostats/pillars!! So they seemed to be transforming the energy into chi and prana, maybe orgone. I also found two iron nuggets that were in the roof on the inside of the cairn, (metal is also one of the components needed to make an orgone generator together with crystals like quartz; most of the stones of the cairns are full of quartz particles).

I somehow felt some kind of odd goosebumps and checked the number of Bovis-units inside the cairn; 74000! (for more information on the Bovis scale; see glossary page 95) The highest Swiss Geobiologist Blanche Merz found was 18000 Bovis units at two different sites on the Tibetan plateau. I wonder whether the number of Bovis units changes inside the cairn according to the time of year or even the time of day. Today is Samhain after all, a very special day when you know that some of the cairns in Ireland are aligned to the sunrise at Samhain. A lady from the American party joined us inside the cairn G and pointed out that you can sit inside three side chambers within the cairn. She tried all three. I decided to try the second one because the American lady said that this felt like it was filling her up with energy, but when I completely sat down with my back against the back stone I felt overwhelmed, over-energised. I felt my back trembling. I quickly got out of the side chamber and the American lady told me that my back was giving off heat. I felt like I was somehow phasing into pure energy and felt too overwhelmed. This was an incredible experience; you could hear the wind howling outside whilst we were safely tucked inside the cairn and it felt like I had warped back in time to the time of the builders. It was clear to all three of us that these weren't burial chambers, but a place of very special energy, maybe even healing energy.

The spot just outside the entrance radiates chi and prana as well and has a much more bearable 9000 units on the Bovis scale. It would be advisable to stand outside of the entrance for a while to benefit from the chi/prana/orgone energy and to not go inside the cairn and get overwhelmed like I did. I was very thankful for Jackie being there and looking out for me.

Picture: Carrowkeel cairn G inside

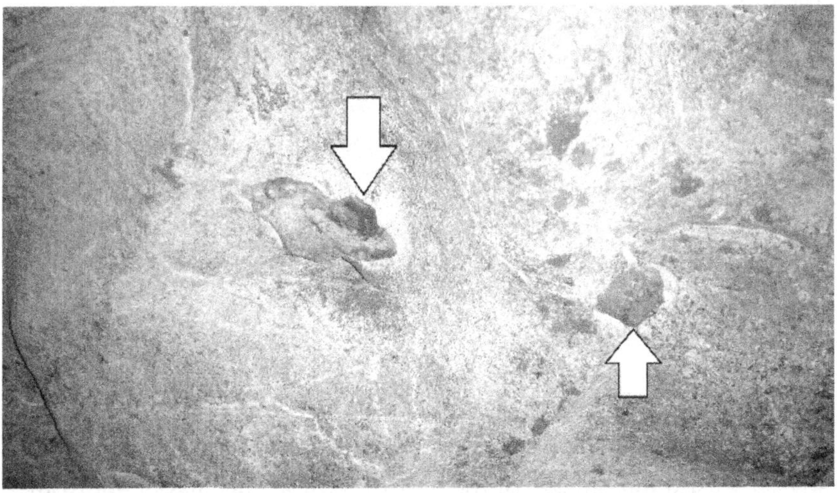

Picture: two iron nuggets in roof-slab at Carrowkeel

Earlier Lecher antenna measurements at the Carrowkeel cairns revealed the 16.4 Leyline and a 12.0 cosmic energy line coming from the entrances of the cairns. Two of the cairns (G and K) entrances are directed 310-315 degrees NW which would indicate the direction of Maeve's Cairn on Knocknarea. The entrance of cairn H seems to be directed towards Knocknashee.

Pictures: some of the cairns at Carrowkeel

CARROWKEEL

by Anne-Marie Delmotte

Carrowkeel
So unreal
The landscape
So unseen
The air so clean
So pure
That angel dust grows on the hawthorn tree
The faerie tree
Covered in mossy lightest of green
Carrowkeel
So unreal
The energies
Transforming
Healing
Universal life force
Prana or chi
Filling me
Transforming me
Into pure energy
Carrowkeel
So unreal
Out of time and out of space
Eden on earth
So blessed am I to be
To have found thee

1.15. MAEVE'S CAIRN ON KNOCKNAREA SLIGO IRELAND AND THE DISCOVERY OF THE 16.4 LEYLINE

Overlooking the Carrowmore megalithic cemetery on the summit of Knocknarea mountain is the huge flat-topped cairn called Maeve's Cairn. 55 meters in diameter and 10 meters high. There are spectacular views from here and the cairn is visible for miles around. It is unexcavated, but it may cover a passage tomb, possibly dating to 3000BC. The name of the monument along with folklore suggests it was built for the mythical Maeve, the Iron Age Queen of Connacht.

Picture: Maeve's cairn as seen from the east

I remember doing some Lecher antenna readings around Maeve's Cairn at the usual settings for Leylines (11.1 and 12.2), but I did not get any signal so I decided to give up on Leylines. I did however find a crossing of two wide cosmic energy-lines at 12.0. I posted my findings in David Cowan's Facebook group Dowsing and Earth Energies. Rory

Macquisten, who is also a member of David's group, and is a dowser and writer, pointed out to me that he did find a crossing of Leylines on Maeve's Cairn and that I should go back and do further readings.

I had also found a signal on 17.6 before which is known as a general dowsing frequency. I tried to find out more and started checking many different settings on the Lecher antenna. I finally found my signal on the Lecher antenna setting of 16.4. There is indeed a crossing of these 16.4 lines where the cairn is. The North-South one is very wide and the width varies. The widest I measured up until now towards the North is 75 degrees and 60 degrees towards the South. The other line is less wide, no more than 1 meter and this one runs NW and SE towards the Listoghil Cairn at Carrowmore Sligo.

Picture: Maeve's Cairn as seen from the south

1.16. ORGONE AT LISTOGHIL CAIRN AT CARROWMORE SLIGO IRELAND

Picture: Tomb 51: central chamber within the Listoghil cairn

After my findings at Carrowkeel I felt that I had to go back to the Listoghil Cairn at Carrowmore megalithic cemetery to check the transformation energy at 18.00 again.

I did indeed find, similar to my measurements in Cairn G at Carrowkeel that all the large stones holding up the roof slab give a signal for transformation (18.00), this being 6 signals in total. I already found a signal for prana and chi and sacred energy here earlier (see 1.12.), so it is obvious that energy is being transformed here too into good energy, beneficial energy, same as at Carrowkeel, maybe energy similar to Wilhelm Reich's orgone (see also 1.12.). Well, at least the same settings on the Lecher antenna are found as with the orgone generators I bought. What an amazing find!

ORGONE

By Anne-Marie Delmotte

Orgone
Who could have known
That I would find you here
Near my new-found home
Ireland
Or was it I that was drawn?
That has always known
You to be here
Wanting to be near
Feeling your pull
To my real home
Where I belong
Amidst constructions of ancient stone

CHAPTER 2: THE LECHER ANTENNA GEOBIOLOGY EXPERIMENTS

For more information about the Hartmann and Curry lines; see chapter 4 on geobiology; grids of the earth.

2.1. STONE CIRCLE IN GARDEN IN IRELAND

For some reason I can't explain I felt that I wanted to make a stone circle on the grassplot in my back garden.

I collected stones from several beaches in Ireland and even brought a few small ones over from the French coast.

I decided to check the energies before and after arranging the stone circle, so I looked for a spot that had a Hartmann crossing and a Curry line running through it. As it happens there was also a 16.4 Leyline running through the spot I chose.

After putting in the stone circle I rechecked the energies and now the Hartmann and the Curry lines were heaped up just outside the circle. It looks like I am on to something here. The 16.4 Leyline on the other hand did not change position and ran straight through the stone circle.

After a lot of researching, I finally came across a reference to a book written by Swiss building Engineer and Geobiologist, Blanche Merz, "Points of Cosmic Energy". She only researched the Hartmann lines, but, like me, did find similar readings of Hartmann lines heaping up like a wall of energy at several special sites.

A similar wall of energy consisting of Hartmann and Curry lines can be found at many different megalithic sites where I did Lecher antenna readings as you can read in this book. It is as if the stone circle creates a space that is clear of negative energies.

Picture: stone circle in my garden

2.2. THE EXPERIMENT WITH THE 4 QUARTZ-CONTAINING STONES IN LIVING ROOM IN IRELAND

Another experiment I conducted was in my living room in Ireland with four quartz-containing stones.

Before: I measured the surface of my living room for Hartmann and Curry lines, a signal for water and for faults. The result can be seen in the following picture. The position where two Curry and two Hartmann lines cross (bucket) is a spot where you wouldn't want to spend much time, as it is a concentration of negative energies.

Picture: Hartmann and Curry lines. Crossing of 2 Hartmann and 2 Curry lines where bucket is placed

For some reason I decided to put 4 quartz-containing stones (arrows in picture) I found at Lissadell beach in Sligo in a rectangle and measured the energies again with my Lecher antenna.

The Hartmann and Curry lines had shifted to just outside the rectangle where they are heaped up, same as with the stone circle in my garden.

In the rectangle two beneficial Leylines (11.1), two golden ratio lines (1.618) and two Schumann-resonance lines (checked at 16.4) had formed. These positive energy lines had all formed in the same place which seemed to be the "middle lines" of the rectangle and they all crossed in the middle of the rectangle. I checked the middle with my labradorite pendulum and it started spiraling outwards in a clockwise direction very vigorously. It seemed a vortex had formed too.

The grounding energy (8.0) that is normally found in the corners of the room at floor-level had shifted to the location of the four stones and the cosmic energy (12.0) of the corners of the ceiling had also shifted to the location above the stones at ceiling-level.

Picture: the four quartz-containing stones from Lissadell beach; ruler in centimeter

The WiFi signal was still operational in the rectangle.

I set out to see whether four quartz-containing stones could shift the Hartmann grid, like in my garden, and what would happen to the WiFi signal - but it seems that I have discovered much more. It looks like this is creating a kind of 3-dimensional sacred space free of negative energy which is replaced by positive energy.

2.3. THE EXPERIMENT WITH THE FOUR ROUGH WHITE PURE QUARTZ-CONTAINING-STONES IN LIVING ROOM IN IRELAND

After the experiment with the four quartz-containing stones I set out to acquire pure quartz stones and see if the quartz in the stones were making these changes to the energies as in 2.2.. I found four small rough white pure quartz crystals at a shop called Tír na nÓg in Sligo. In Irish mythology and folklore, Tír na nÓg ("Land of the Young") is one of the names for the Otherworld. It is depicted as a supernatural realm of everlasting youth, beauty, health, abundance and joy. The shop sells all kinds of organic products.

I repeated my experiment and found the exact same result as with the four quartz-containing stones in 2.2..

See 2.5.-4 for a picture and size of these four rough white quartz-containing stones.

2.4. THE EXPERIMENT WITH SEVERAL CRYSTALS IN LIVING ROOM IN BELGIUM

As I recently acquired some new crystals, I was eager to test them so I did an experiment with these crystals in the living room of my flat in Belgium. As you read earlier in this book, I had a good result in neutralising a crossing of Hartmann and Curry lines with amethyst and rose quartz (see 1.5.) and wanted to see whether this can be reproduced.

My flat in Belgium is very small so I was quite limited in what I could check.

I found a crossing of two 1.7 grid lines that came together with one Hartmann line.

I checked for the effect of following crystals: amethyst, quartz-amethyst, lapis lazuli, red agate, red jasper, green jasper, malachite, galena, shungite and labradorite.

These are my findings:

-the crystal has to be placed with precision as the 1.7 lines are only 1 centimeter wide

Amethyst, quartz amethyst, lapis lazuli, red (more brownish) agate, galena and shungite neutralise the lines completely. Some of the crystals are quite small, a larger one would be more advisable.

Green jasper, malachite, red jasper and labradorite have no effect on the lines.

I did all my readings with the Lecher antenna in positive (magnetised rod in left arm in positive mode) except for the verification of the three lines that crossed where I verified in negative to see whether the signals were genuine (see glossary page 95).

2.5. THE EXPERIMENTS IN KITCHEN IN IRELAND

I wanted to reproduce some of my earlier findings (see above in this chapter) at another location in my house, this time my kitchen, and by using single crystals and various sets of stones and crystals.

I scanned with my Lecher antenna for Hartmann and Curry lines; energy of running water and a Leyline at 16.4 first. I found a particular spot that was ideal for testing the influence of the stones and crystals that has a Curry line crossing and a Hartmann line crossing and that also has a running water signal going through it that is coming from a waterpipe that comes from the mains in the street and going to my kitchen sink. The spot also has a 16.4 Leyline running through it.

I checked the influence of the following setups:

1. Setups

A. I put the four pure quartz stones I used earlier in my experiment in my living room in all four corners of my kitchen.

I measured and there were no changes. I found this somewhat odd, but then I realised that the crystals were not put in an exact rectangle. One was in front of the fitted kitchen furniture.

B. I moved the crystals into a perfect rectangle and measured again.

All signals were gone: the Hartmann, Curry and water current signals were now negative. Only the 16.4 Leyline was running through the space as before.

Two Leylines at 11.1; two sacred energy lines at 1.1; two Schumann resonance lines and two golden ratio lines at 1.618 had formed, all crossing on the same spot that looked like the middle point of the rectangle. Also, a clockwise turning vortex was formed at this middle point (checked with pendulum).

I asked, with the antenna, whether it was beneficial to be in such a space and the answer was "yes". The answer to the question of whether the crystals need regeneration, was "no".

The space was 12-13000 Bovis units which corresponds to a space of spiritual quality (10-14000 Bovis units).

The Hartmann and Curry lines had all shifted to just outside the crystals where they were all heaped up - similar to my findings at cairns and stone circles.

It looked like the four quartz crystals were able to clear all negative energy from a space and bring in positive energy.

The energy configuration went back to its original configuration when the crystals were removed.

I realised that where the Hartmann and Curry lines were heaped just behind the crystals were actually the contents of my fridge, hot water kettle, and various other objects I'd rather not have being exposed to a pile of negative energy. See point C. for solution.

C. I put two of the crystals in the corner on the work surfaces of the fitted kitchen in order to create an exact rectangle albeit a three-dimensional one and did my readings again. This did not influence the outcome. It was the same as in 1.B. Apparently, you only have to make sure that you have a perfect rectangle, that can also be three-dimensional. This is a much better configuration, because at least now the negative energy lines are heaped up outside of the kitchen space and not influencing food or other products.

It looks like negative energies were cleared and positive, beneficial ones were put in. However, questions spring to mind. The place feels very "clean" and it's like I'm in an open space. It's really great to be in such a space, but it took me a very long time to reprogram myself to not feel hindered by negative energies. Will staying in this kind of space for a long time lower my defenses?

Another fact is that my house stands on very special land amidst more than 5000 years old megaliths that were put here for a reason. My house stands on two different alignments, each between two major cairns of this region. Can this result be reproduced elsewhere? Is it the site? Is it the practitioner-architect? Blanche Merz says it is both as she demonstrates in her book with large structures in Egypt having been moved to another site that completely lost their energies.

I think the created negative energy free environment and positive energies are perfect for meditation, writing, painting or just chilling out, but to live in .... This remains to be experienced.

A friend of mine who also has a Lecher antenna was so kind as to test four quartz-containing stones I tested in my kitchen before, at her house and she found similar results to mine. This could mean that the site or the architect might not be that important, although further investigation needs to be done before coming to any definite conclusion.

It is now more than one year later and after having conducted research in my house and at several other habitats and places, and of the Megaliths in Carnac and Brittany France which can be found in my newer book called "Signpost to the Holy Grail? Infinite Energy with Infinite Possibilities! Dowsed with the Lecher Antenna in Carnac and Brittany France", I like to add following information to the second edition of this book:

"I have put my own house and gardens in a very special energy configuration by using materials I found in Ireland and more recently also in Brittany where there are Megalithic Monuments. This configuration creates a protective shield around my house and gardens - first floor included - which shifts out/clears all known geopathic stress (see chapter 4) to just outside the house and gardens. This configuration brings in positive energies to every room in the house, even the hallway, these positive energies being sacred and divine energy, Schumann resonance, positive Leylines, portals to a higher vibration, a clockwise turning vortex, golden ratio, ... I also found a way to neutralise the build-up negative energies around the configuration too so that the neighbours are not affected (in case of apartments, flats, terraced houses, ...).

I found out that this energy configuration is similar to that of the Quadrilatères - rectangular Megalithic Monuments - of Crucuno and Manio near and in Carnac in Brittany France.

The corners of the Quadrilatère of Crucuno are aligned to the summer and winter solstice sunrises and sunsets and its dimensions are apparently according to the Golden Ratio.

I am able to create this energy configuration no matter the angle and no matter the dimensions. I did this with my house by means of an experiment, but I left my house in this configuration because it just feels very good. I have the impression of being in open air and in a very clean place without it feeling sterile. It has now been over a year since I put in this configuration. It also feels safe, because I found out/measured that even other kinds of negative energies and bad thoughts/thought-forms like jealousy, envy, ... cannot penetrate the protective shield.

I am currently working on a more elaborate course in geobiology I hope to offer in the future and am looking at including how to put a habitat in the very special energy configuration of the Quadrilatères."

2. I then checked whether I was able to correct the spot with the Curry-line crossing and Hartmann-line crossing and running water signal with a single crystal. I tried this with a rough amethyst crystal. I put the amethyst on this spot. This resulted in the two Curry lines as well as the two Hartmann lines that formed a crossing at this spot dissipating. The water signal also was gone at the height of the correction with the rough amethyst.

The 16.4 Leyline was running through the corrected spot as before.

The energy configuration went back to its original position once the amethyst was removed. See 4. for picture of the rough amethyst.

3. I tried the same as in 1.B. above, but with four black tourmaline crystals. There was no change to any of the energies. See 4. for a picture of the four black tourmaline crystals.

4. I tried one black tourmaline crystal the same way as the rough amethyst in 2. above and obtained the same results as in 2.

Picture: crystals used in the experiments:
first row: four rough white pure quartz stones
second row: four black tourmaline crystals
third row: rough amethyst; ruler in centimeter

5. I tried an orgone generator the same way as the rough amethyst in 2. above and found there to be no immediate change to the energy-lines.

6. I repeated the experiment as in 1. C. above, but with the four quartz-containing stones I used earlier in my experiment in my living room (see 2.2. for picture) in all four corners of my kitchen and obtained the same results as in 1.C. above, except that the positive energy that was formed felt somewhat weaker. 11-12000 Bovis units. These stones require no maintenance.

7. I repeated the experiment as in 1.C. above, but now with four large quartz-containing stones I found earlier this week at Strandhill beach in order to check reproducibility. I put them in all four corners of my kitchen and obtained the same results as in 1.C. above, except that the positive energy that was formed felt somewhat different. 11-12000 Bovis units. These stones also need no maintenance.

Picture: the four large quartz-containing stones from Strandhill beach; ruler in centimeter

8. I repeated the experiment as in 1.C. above, but now with four medium sized quartz-containing stones I found earlier this week at Strandhill beach in order to check further reproducibility. I put them in all four corners of my kitchen and obtained the same results as in 1.C. above, except that the vortex that was formed was in a slightly different spot and the Bovis units of this set up are slightly higher; 12-13000. These stones also need no maintenance.

Picture: the four medium sized quartz-containing stones from Strandhill beach; ruler in centimeter

9. I was able to buy four rough rose quartz stones from the same shop Tír na nÓg in Sligo. I checked these the same way as in 1.C. above. The obtained result was the same as in 1.C. above.

Picture: the four rose quartz crystals; ruler in centimeter

2.6. EXPERIMENT IN FLAT IN BELGIUM: HOW TO PROTECT YOURSELF FROM NEGATIVE ENERGY AND NOISE FROM YOUR NEIGHBOURS

I had an issue recently with temporary neighbours and asked fellow Lecher antenna users and dowsers for help. I felt that the neighbour(s) were draining my energy and found a signal for a very negative energy coming from their flat.

Someone advised me to work on a white piece of paper on which I draw the common wall with the neighbours in pencil and seal off this wall on the paper with copper wire and then fill the wall with white stones until there are enough stones for there to be no longer a signal of negative energy coming from the neighbours.

I took this one step further and actually sealed off the wall as such by using the Lecher antenna set on the frequency of copper. I did this all along the wall in a clockwise direction with paying special attention to the corners and by using a dynamic sending technique.

After this I filled the walls with the clear quartz frequency by using the same dynamic sending technique with the Lecher antenna until the antenna stopped moving meaning the walls were sufficiently filled. More sending of this frequency was needed where there were electrical wires, fuses or electrical equipment (TV, ...) and also a wide Curry line that shows a significant spiral signal (see glossary page 95).

The earlier measured very negative energy from the neighbours was now no longer measurable so I would tend to assume that it worked.

The reading I did was very much against my principles (of a person without permission), but I found this to be a case of force majeure. One of the temporary neighbours was draining on my life energies which is very unusual for me as I try to protect myself as much as I can.

I also set up my antenna to broadcast me 13.00 for my protection as this has helped me in the past. 13.00 is the frequency of white light.

THE MAGICAL STONES

by Anne-Marie Delmotte

The wild Atlantic
The refreshing smell of seaweed
Full of iodine
All of this, mine
For this swift moment in time

Soft sand and stones
Holding relics of ancient bones
Stones and stones and stones
Their tiny quartz particles glittering in the sun
Like sapphires
So beautiful
So powerful

Who could have known
Their power to conjure up
Sacred energies
Create sacred space
In any odd place
Thank you crystals
For your magical gifts

CHAPTER 3: BIO-ENERGY METHODS

3.1. THE DELMOTTE VIBRATING ENERGIES RÉSONNANTES METHOD OR DELMOTVIBRES-METHOD

This method is based on several Lecher antenna methods I have studied and my experience with them, but also on my many Reiki trainings and experience, and on Traditional Chinese Medicine.

Sources:

-Online e-learning course Dowsing with the Lecher Antenna – Bio-energy Basic and Intermediate Level by Dame Anne-Marie Delmotte

-Practical Guide for Dowsing with the Lecher Antenna – Elaborate Basic Training Course in Geobiology and Bio-energy by Dame Anne-Marie Delmotte

Both online course and Practical Guide come with a Balance Sheet, a drawing of the positions of the lymphatics and the chakras and a Bovis Scale that can be used to measure the person's vibrational quality before and after their balancing to see how well they improved.

Balancings can be done on yourself or on another person. The course is meant for anyone, even those who have no experience at all with the Lecher antenna because it covers how to prepare and hold the Lecher antenna as well. The online course also shows what physical exercises can be done to learn to familiarise yourself with the antenna dipping (signal).

There is an explanation how distance work can be done and the course is open to the use of different Lecher antennae.

The online course consists of short one to one videos, each addressing one subject of the course. The course further contains spoken tutorials supported by presentations. The aim of this course is to break down the training into very simple step by step instructions and easy to perform exercises.

The protocol covers the following areas:

READING OF THE BIO-ENERGIES

The person's vibrational quality is first measured on the Bovis scale.

After this, the person's bio-energies are measured to see which signal(s) are absent or off centre and to be able to make a comparison before and after the balancing.

Following signals are measured:

-cosmic connection at 12 at the height of the crown chakra

-telluric connection at 8; the signal is checked and also how deeply the person is rooted into the earth (vertical scan)

-the 15.3 is measured at the chest level which is the synergy between the 8.0 and the 12.0 and to see how centred this signal is (vertical scan)

-the 17.6 is measured at the chest level which is the ability to eliminate debris of light photons

-the 15.3 is read on the forehead which is the combination of the magnetic and electrical field

-the 17.6 on the forehead is checked for balance

-the 1.1 is measured at the chest level. 1.1 is the bridge between the outer centrifugal and inner centripetal field. It is the gateway or bridge to the deeper energy-reserves (ACMOS discovery).

-the lymphatics are read at 3 levels left and right at 2.1

-the 5 chakras are read for their vibrational signal at 2.5

-polarity check: position of head=negative polarity and feet=positive polarity

-the magnetic field axes are read frontally and laterally

-the electrical field axes are read frontally and laterally

-the vibrational field axes are read frontally and laterally

CHECK OBJECTS AND MOBILE PHONE WORN BY THE PERSON

After this initial reading, the person is checked if objects and mobile phone worn on their body have an influence on these readings. The objects will be made compatible if necessary and/or possible.

The influence of the mobile phone is checked and rechecked after use of a phone corrector patch or wearing a certain crystal on the body. It is also taught how to look for the spot on the mobile phone where to glue the corrector patch.

If some readings are still closed or off centre, further balancing is done with the DELMOTVIBRES-Method.

BALANCING OF THE BIO-ENERGIES WITH THE DELMOTVIBRES-METHOD

The DELMOTVIBRES-Method consists of:

-working on the **meridians**; each meridian has one or more frequencies they should be on to properly function. The meridians are measured and checked for the acupoint(s) responsible that is/are disturbing the meridian(s) if the case. The different acupoints of the meridians are related to many different physiological and anatomical pathologies: organs, mental state, ...

A deep understanding/extensive study of the meridians or Traditional Chinese Medicine is not compulsory as the energetic body of the person will guide you to what it wants you to do. In the course references are made to where you can find simple and detailed information on acupoints which can easily be accessed (internet, book) and also do not require extensive studying but that will give you a better understanding of what is being balanced and what indications are being addressed.

-working with **colour harmonies**; these work on a multitude of pathologies too. During a certain balancing, I even had to conclude that it worked on a back-pain!

-working on the **chakras**; each chakra has several series of one or more frequencies they should be on to properly function. They are corrected after asking the energetic body for the explicit permission to do so. The chakras, like the meridians, are related to a whole series of pathologies but extensive knowledge is not necessary either as the energetic body knows what harmonies are available and will choose them accordingly. And as with Reiki; the energy will go where it is needed. The online course does however explain what each chakra is related to.

Working on the meridians, the chakras and with the colours encompasses a lot of harmonic frequencies that have an influence on different levels (organs, nervous system, emotional, spiritual, psychological, ...). The energetic body has many possibilities to choose from to balance itself.

The most important factor to me is that you approach the person with the pure intention of wanting to help and with love. The Lecher antenna, the harmonies and the energetic body of the person will guide you through the process of the balancing. You will be amazed at the outcome. I certainly am, time after time!

-the protocol is then worked again to see whether further corrections are needed

The person is checked after the balancing for their vibration on the Bovis scale and the signals which were closed or off centre are rechecked.

If you would like to learn more and be able to do this yourself I would advise you to take the online course: "Dowsing with the Lecher Antenna Bio-energy Basic and Intermediate Level" which is available on udemy.com or you can self-learn by means of the Practical Guide: "Practical Guide for Dowsing with the Lecher Antenna – Elaborate Basic Training Course in Geobiology and Bio-energy".

Only a small investment is needed as you only need a Lecher antenna and the optional colour filter harmonies which I would highly recommend as they are very effective. I make these sets myself which I can sell to my students to keep it even more low budget.

My aim is to make working with the Lecher antenna available to as many people as possible.

Something which is highly underestimated is the influence of the habitat and it is very important not to sleep on negative energy or geopathic stress (see chapter 4). The Lecher antenna also allows to measure the homeplace, in open air (crops) and other sites (like where cattle is kept) in order to look for geopathic stress and to find out whether a person's bed has to be moved or corrections are needed. How to measure this together with ways to improve and enhance the quality of rooms, offices and spaces in general is taught in the geobiology part of the "Practical Guide for Dowsing with the Lecher Antenna – Elaborate Basic Training Course in Geobiology and Bio-energy" and there is also a specific online course called "Dowsing with the Lecher Antenna Geobiology Basic Level". This course is also available on udemy.com. There is lifetime access to the online courses and there is a questions/answers forum where you can communicate with the instructor, yours truly and other students. I am very honoured and pleased to say that also doctors and physiotherapists from all over the world, even China are taking my course and who are very satisfied as they can immediately put it into practice and they also have let me know that it's proven and proving to be very effective. You do although not have to be a doctor or a physiotherapist to work with this method as I made it accessible and easy for everyone to use.

My hope is for the DELMOTVIBRES-Method to be recognised by insurance companies as a standalone method, meaning to everyone who wants to put it into practice, in as many countries as possible in which I already am succeeding, but there is, still a lot of work left to be done.

There are vouchers at the end of this book which give you a 20% discount for the online bio-energy and geobiology course.

3.2. THE ACMOS METHOD

Source: ACMOS training course level 1 and 2 given by Carol Robertson in Scotland September 2017

I learned so much during this training course that it is not easy to summarise the method on a few pages.

The protocol is as follows:

The person's energies are measured first extensively in order to be able to make a comparison of these readings before and after the balancing.

The reading can include the measurements of the Ting-points on hand(s) and foot/feet if the energetician has a Biofeedback instrument. This allows an objective comparison of the meridians before and after the balancing.

Reading with the Lecher antenna consists of:

-reading of the 8.0 (telluric or earth energy) below the centre of the pelvis and the 12.0 (cosmic energy) at the top of the head

-the 5 chakras are read at 2.5

-the lymphatics are read at 3 levels left and right at 2.1

-the 15.3 is measured at the chest level which is the synergy between the 8.0 and the 12.0 (scan from left to right)

-the 17.6 is measured at the chest level which is the ability to eliminate debris of light photons

-the 15.3 is read on the forehead which is the combination of the magnetic and electrical field

-the 17.6 on the forehead is checked for 5 different balances; vibratory, organic, mental, hormonal, lymphatic. The ACMOgraph is better suited for these readings.

-the magnetic and electrical field can be read on the person and/or by using the ACMOgraph frontally and laterally. Both ways of measuring give the energetician different information.

-the 1.1 is measured at the chest level. 1.1 was discovered by ACMOS and is the bridge between the outer centrifugal and inner centripetal field. It is the gateway or bridge to the energy-reserves.

-The ACMOgraph gives additional information which is obtained by using a general dowsing setting on the antenna.

After this initial reading, the person is checked if objects worn have an influence on these readings. The objects will be made compatible if necessary and/or possible.

If some readings are still closed or off centre further balancing is done with the full ACMOprotocol.

The energetic body of the person has the choice of following harmonisations: ACMOS protocol, Pathos, Organic Pathos, Stimulation Protocol and Environment.

- In case **ACMOS** is dominant vital energy resources are prioritised and should be brought to optimum level. In this case a harmony is selected by the energetic body.

There are many different harmonies: essential oils, minerals, trace-elements, homeopathics, herbs, seasonals, ...

Once a harmony is selected, an application method is chosen by the person's energetic body from the following possibilities: acmodynamise, harmonise, reinforce, pathos, place, follow up and inhale.

- In case **PATHOS** is dominant, mental or organic disorders have priority and necessary medication should be given to the subject when ACMOS energy is restored. In many cases, the patient requires orthodox or alternative treatment first because the pathology goes beyond energy disorganisation.

- In case **ORGANIC PATHOS** is dominant there is a neglected element requiring reintegration of the body's energy system in order to achieve harmony. In this situation, there is a problem or disorder of one of the elements – an organ, gland or system – which has caused excessive imbalance to the global energy harmony, causing the global system to abandon or isolate the element. With the ACMOS method, global energy will accept reintegration of the abandoned element to fully restore global body energetic organisation.

- In case **STIMULATION** is dominant, the body requires treatment relative to the acupoints to restore correct polarity and functioning as necessary for global energy balancing.

A method of stimulation is selected from what is available to the practitioner: Quantacmos (lightpen), Lecher antenna, or electronic (Acmodermil).

Stimulation points or pairs are selected from the ACMOS STIMULATION PROTOCOL that are arranged in order of priority.

- In case **ENVIRONMENT** is dominant, the environmental protocol is consulted. The origin (geopathic, ecological and ambient) is determined in case of a leak and the environment aggression is identified if there is a leak or not. An environmental aggression can be one of the following:

Relational (mental or physical), human energetic residue, heavy metal (amalgam, deodorant), radio-activity, mobile phone, television set, power cables, micro-waves, medicinal, nutritional, vaccinal (personal, parental), virtual, space acmogram, personal acmogram.

Often an essential oil and/or colour filter is/are needed, but the oil can, in some cases, be replaced by an acmogram, if placement of the oil responds. This would be in the case of a brain confusion; the aggression is of a virtual nature.

This brief summary only gives a short overview of what is possible with the ACMOS method and is certainly not exhaustive. At this point I did the level 1 and 2 training courses.

3.3. GENERAL CONCLUSION ON BIO-ENERGY METHODS

My first idea was to compare the above mentioned methods, but I have to admit that they are both so effective that I don't feel like going there at all. To me both methods are valid and demonstrate good and effective results.

But the more you know, the more you can do. I am, at this point, not in possession of all harmonies and equipment for the ACMOS method because it is a large investment.

For the Delmotte Vibrating Energies Résonnantes Method or DELMOTVIBRES-Method only a Lecher antenna is needed, even the colour filter harmonies are optional which means that with only a very small investment a lot of "issues" can be addressed.

But for what I do have and what I can do, I leave it up to the energetic body of the person being balanced to choose what it prefers to have done.

There is **another very different bio-energy method** that was developed on the continent by a person who worked with Dr. Hartmann (from the Hartmann grid of the earth; see chapter 4) and Reinhard Schneider, who developed the Lecher antenna, that I will definitely look into as well. I met a retired medical doctor very recently who worked on about 10.000 patients with this method who has given me further information and plenty of case studies. I mention two of his cases in chapter 5.

CHAPTER 4: GEOBIOLOGY

4.1. GEOPATHIC STRESS AND HEALTH

Geopathic stress is a form of trauma caused by disturbed or anomalous energies within the earth's mantle. The earth is surrounded by energy grids that can become harmful to life.

Geopathic stress has been implicated in a number of undesirable effects to human health, from simple conditions such as sleeplessness or confusion to highly dangerous ones such as cancer, decreased fertility, and auto-immune dysfunction.

There are, besides the grids of the earth, many other causes of geopathic stress. It is very important not to sleep on geopathic stress. The resistance of the skin drops considerably when we sleep, which makes us even more susceptible to geopathic stress.

Geopathic stress can also be measured on a person. This gives a good indication of what the energy in and around his/her bedroom looks like to which the person is exposed to when sleeping. The energy of the immediate surroundings leaves a kind of an imprint on the person this way. I have recently had a meeting with a retired medical doctor who has treated about 10.000 patients by using this method who came to him because they kept on being ill and were desperate because they could not be helped with orthodox medicine. The influence of the energies a person's habitat is very important and often not addressed by orthodox medicine at the moment. Only very few medical doctors are aware of this and use the Lecher antenna in their practice.

4.2 GRIDS OF THE EARTH

Lecher Antenna settings are for Lecher antennae with a 20,05 centimeter ruler (most commonly used ones). See also chapter 6.

The earth is surrounded by several energy grids. These are found everywhere (in dwellings or on open ground) and are regular, although their mesh can sometimes be distorted by interference. The first two grids are called after the person who discovered them. They have following characteristics.

1. HARTMANN GRID

Lecher antenna setting=12.2
They run North to South and are 2 meters apart.
The ones that run from East to West are 2.5 meters apart.
Their width is 21 centimeters.

2. CURRY GRID

Lecher antenna setting=6.9
They run Northeast to Southwest and are 2.44 to 3.51meters apart.
The ones that run from the Northwest to the Southeast are also 2.44 to 3.51 meters apart.
Their width is 30.5 to 40.6 centimeters.

3. GREAT GLOBAL GRID

Lecher antenna setting=7.6
They run North to South and are 24 meters apart.
The ones that run from East to West are 30 meters apart.
Their width is several dozens of centimeters.

4. *GREAT DIAGONAL GRID*

Lecher antenna setting= 8.2

They run Northeast to Southwest and are 36 meters apart.
The ones that run from the Northwest to the Southeast are also 36 meters apart.
Their width is 3 centimeters.

5. *1.7 GRID*

Lecher antenna setting=1.7

They run North to South and are 2.5 meters apart.
The ones that run from East to West are also 2.5 meters apart.
Their width is 1 centimeter.

More information on the other causes of geopathic stress and how to measure and make certain corrections can be found in the "Practical Guide for Dowsing with the Lecher Antenna – Elaborate Basic Training Course in Geobiology and Bio-energy" or online course "Dowsing with the Lecher Antenna Geobiology Basic Level" available from udemy.com. There is a voucher at the end of this book which gives you a 20% discount for the online geobiology course.

CHAPTER 5: BIO-ENERGY AND GEOBIOLOGY CASES

The balancing of energies for some of these cases was performed by means of distance work. Distance is no issue. I balanced persons that are as far as 8130 kilometers away.

5.1. CASE: MY MOTHER - DYING

Pathology: dying; incurable Clostridium difficile diarrhea

Major corrections during balancing: this was in 2012 and I only had basic training so I tried to work with what I knew at that point in time. The results were so amazing and eye opening that this was my trigger to do further bio-energy training with the Lecher antenna.

I mainly worked with broadcasting (leaving the antenna to send a certain frequency in distance balancing) for several hours and/or overnight.

I sent my mother 17.6 for regeneration; 8.0 for grounding as this energy was very weak indicating that she was about to leave us. This had definitely helped, because she made a miraculous recovery as she was supposed to only live for one or two more days. A few days after I sent 3.75 to strengthen her immune system and 4.15 to strengthen her nervous system (for her dementia). In just over two weeks she went from dying to being cured from her incurable Clostridium difficile diarrhea.

Result: **she lived for another three years after this.**

5.2. CASE: MY STEPDAD – NOT RECUPERATING AFTER OPERATION

This is a geobiology case I find well worth mentioning.

My stepdad had a prostate operation many years ago and he just wouldn't recover from this. It had been 6 weeks since the operation and no improvement. So I told him that his hospital bed might be placed on a crossing of energy lines or maybe on a crossing of a water vein with other energy lines. I asked whether it was okay with him to check this to which he agreed.
So I used my Lecher antenna and scanned for Hartmann and Curry lines and water veins and faults. As it seems on the very spot he was operated on there was indeed a crossing of several energy lines. So I moved the bed some 20 centimeters and informed the staff of the latter so they wouldn't move the bed back to its original position.

<u>Result:</u> **he had a speedy recovery after this and left the hospital shortly after**.

5.3. CASE: FLUE LIKE SYMPTOMS AND FATIGUE

I did a balancing on myself which is a very good example of how effective the <u>DELMOTVIBRES-Method</u> is.

<u>Pathology</u>: flue like symptoms and fatigue

Energy readings show following signals being weak or not good:

- 1.1 signal on the heart chakra is off; this is the bridge to the energy reserves; this means that I have no access to these reserves and am exhausting myself because of this.

- 15.3 signal on the forehead is way off the left as are the axes of the electric field and magnetic field; the 15.3 on the forehead is the combination of my electric and magnetic field.

A frontal displacement of the axes means there is physical cause for this.

-Two signals of my lymphatics are weak on my left side. The lymphatics help us eliminate toxins from the body.

-All chakras show a vibrational signal, although the base chakra has no signal for one if its right frequencies. I ask my energetic body for permission to bring it back to the correct frequency which I get.

I bring the base chakra back to its frequency. This has opened the 1.1 bridge to my energy reserves and brings the axes of my electric and magnetic field back to normal.

-I then check for colour harmonies. I get a signal for 1 colour for 4 minutes. This brings the weak lymphatic signals back to normal.

Result: **The flue like symptoms and fatigue are gone immediately after the chakra is put back to one of its right frequencies.**

5.4. CASE - SEVERE SHOULDER PAIN

Method used: DELMOTVIBRES-Method

Symptoms: severe pains in left shoulder since a few days, too painful to lift left arm.

The physiotherapist, who is also an osteopath, is not capable of relieving the pain and advises a bio-energy balancing with the Lecher antenna as the physiotherapist thinks it is stress related. As I already worked on this person before, I start with checking the meridians. One of the meridians is not on one of its right frequencies. This frequency is often associated to shoulder pain, even to shoulder pain blocking all of the lymphatics. I put the meridian back to its missing frequency without working on the other right frequency which the meridian does have and which is associated to another pathway on this meridian.

Result: **the shoulder pain is immediately gone after my intervention and hasn't come back for several months according to the person.**

5.5. CASE - INSOMNIA/EXHAUSTION

Method used: DELMOTVIBRES-Method

Symptoms: person is awake for several hours during the middle of the night with exhaustion as result.
One of the meridians is not on one of its right frequencies. One of the indications for the acupoint responsible for this meridian to be disturbed is associated with insomnia. I put the meridian back to the frequency in question.
The energetic body wants to be balanced with three different colours: colour 8 and 10 which are both associated with energy leaks and colour 20 which works on improving the energy exchange.
The solar plexus chakra is not on one of its frequencies. I correct this after the energetic body of the person has given me permission to do so.
Result: **the person feels much better and is able to sleep without being awake for hours on end during the night. The person no longer feels drained during the day either.**

5.6. CASE – RHEUMATIC PAINS, LOWER BACK PAIN AND FATIGUE

Method used: DELMOTVIBRES-Method

Symptoms: fatigue, drowsy, rheumatic pains and lower back-pain (sciatica)

An energy reading of the person reveals that all signals for the lymphatics are closed which means the person is no longer able to eliminate toxins which are now build up in the body and very likely causing the pain and the drowsiness. The 1.1 signal at the height of the heart is closed too which means that the person has no access to his/her deeper energy reserves and is exhausting him/herself causing the fatigue.

Two meridians are put back to one of their right frequencies and the energetic body wants to be balanced with five different colours.

The solar plexus chakra is not on one of its frequencies. I correct this after the energetic body of the person has given me permission to do so. This opens all lymphatics and the bridge to the energy reserves. The base chakra is not on one of its right frequencies and put back after the energetic body has given me permission to do so.

Going back over the protocol, one of the meridians needs to be corrected again, as does the base chakra. There are two acupoints responsible of which one is the same as during my first correction of the meridians, this being bladder 36 which is associated with sciatica (one of the symptoms was lower back-pain). The base chakra is put back to its right frequency again after permission is given by the energetic body of the person.

I then checked the meridians again and the bladder meridian comes up again, now for acupoint 60. Bladder 60 has a direct influence on the whole bladder channel and reduces inflammation from the whole length of the channel including sciatica, acute and chronic back-pain.

I go over the DELMOTVIBRES protocol again and now no further balancing is required.

Result: **the person is no longer drowsy and no longer has any rheumatic pains. The lower back-pain caused by the sciatica is less and disappears after a few days.**

5.7. BIO-ENERGY CASE: MIGRAINE

Symptoms: migraine
This is actually a quite remarkable case as this just happened.
After giving a lecture, one of the attendees bought one of my books and my Practical Guide and as a few others were very keen on starting to work with the Lecher antenna as well, I started explaining my "Delmotte Vibrating Energies Résonnantes Method or DELMOTVIBRES-Method". I elaborated on how to scan the meridians and I asked the attendee who had just bought my Practical Guide whether I could demonstrate this on her. She told me that she had a lot migraines and actually had one at the very moment. I checked for the position of her head (negative polarity) and measured that it was off to her left. I scanned the meridians by means of demonstration and I got a signal for the governor vessel, meaning this meridian was not on its right frequency. I put it back to its right frequency and measured that this had brought the position of her head back to centre.

After a matter of not even a minute, the person told me her migraine had gone! Amazing!

5.8. REMARKABLE GEOBIOLOGY CASE: MIGRAINE AND WEIGHT LOSS

Symptoms: migraine and weight loss

The person asks me for a house clearing.

After consulting medical doctors and having all kinds of medical tests and exams done, the cause for the migraine and the weight loss cannot be found. The person suspects an energy problem in the habitat.

The place where the person lives is quite special as it is very near to Megalithic Monuments and there is even a standing stone in the person's garden of which the person would like to know whether it is functioning energetically. Right up my alley and a very interesting case for me as I have done extensive research of the Megaliths in Ireland and France after writing the first edition of this book.

The house is most beautiful, it was completely renovated and it is obvious that the person has gone to a lot of trouble of making it very appealing. There is no clutter anywhere and the colours and decoration were meticulously chosen.

The person prepared different plans of the house and garden and its surroundings for me to have a look at and work with.

When looking at the map of the different rooms and after looking for the earth's grids and other causes for geopathic stress I get severe chills when I notice that a toilet is situate where the exact middle of the house is. The toilet in question gives a very strong signal for stagnant water which is enhanced by the fact that it is in the middle of the house (energylines), a most important place for the energy to flow. The house has only 6.000 Bovis units (for more information on the Bovis scale; see glossary page 95) meaning the house is not vibrating as it should. The toilet in question is only at 500 Bovis units.

The standing stone in the garden is fully functioning meaning that there are no energylines in the house of the earth's grids because they are drawn in by the standing stone. More information on the energies of different types of Megalithic Monuments can be found in my book: "Signpost to the Holy Grail? Infinite Energy with Infinite Possibilites! Dowsed with the Lecher Antenna in Carnac and Brittany France".

Further investigation of the bedroom of the person reveal that electrical equipment is plugged in during the night and causing electrostress which is also caused by wiring in the walls and power plugs and switches in the wall. In other words, nothing in this house was earthed. I advise to unplug the equipment during the night and we look for the best spot on the bed to sleep with my HF/LF magnetic field and electrical meters where the electromagnetic radiation is extremely low.

I then work on the toilet with colour harmonies after selecting them with the Lecher antenna which brings the house up to 22.000 Bovis units. I check the person's bio-energies and look for an imprint of geopathic stress on their body as well but all readings are good.

I advise the person to keep the door of the toilet in question open as much as possible and to install a yellow light in order for the energy to flow.

I think it's a wrap and ask if I may use the toilet. The person takes me to the toilet responsible-although there are many other toilets in the house- for the house being sick which turns out to be a blessing in disguise as I sense that the toilet needs further corrections. The signal for the stagnant water needs to be deviated or neutralised. The person is aware of my experiments with the four "Magical Stones" and the special energy configuration they create, see the Experiments in chapter 3 of this book.

The person asks me to put the house into this special energy configuration. I luckily have four stones with me and I configure them in a rectangle around the house. This shifts the stagnant water signal to just outside where the four stones are configured clearing the house of its negative energy (stagnant water signal) and bringing in positive energies into every room of the house.

The house goes up in vibration to 26.000 Bovis units. I now felt that the house was going to be able to keep its higher vibration.

Result: **it is now some time later after the house clearing and the person tells me the migraines have kept away, except for the odd slight headache and that the person feels more energetic than before. The person has also gained some weight since. Other people visiting the house say that it feels "vibrant".**

5.9. CASE: TOOTH PROBLEM?

After meeting with Dr. Driscart who worked on about 10.000 cases with the Lecher antenna and going over his cases in his book in Dutch that no longer available is on the market for purchase I asked him whether I can write about some of his cases here as the information is very interesting. I limit myself to two of his cases which clearly demonstrate the importance of the location of a person's bed.

First case is of a 30 year old man who goes to see the doctor with a pain on the left side of his face.

Previous history of this patient is as follows:

-the previous year he starts to have pain on the left side of his throat. X-ray, echography and an ear-nose and throat examination do not show anything. He suddenly develops an abscess on his left wisdom tooth. He gets treated with antibiotics and the tooth is extracted after which the throat ache on his left side disappears.

One year later the pain is back on the very same spot on his left throat. Again, the dentist, the ear-nose and throat specialist can't find anything and the X-ray does not show anything.

Now the pain has spread to the left part of his face too and no one can help him or tell him what is going on.

So he goes to see Dr. Driscart who works with the Lecher antenna who measures the geopathic stress imprint on his body which reveals that the spot on his throat and the left part of his face are exposed to severe geopathic stress.

The patient made a speedy recovery after this.

5.10. CASE: ASTHMA

Another of Dr. Driscart's cases I would like to mention here is that of a male patient who complains of hay fever, runny nose and eyes, wheezing and heavy chest.

Measurement of the patient reveals that he is sleeping on geopathic stress, one signal being that of running water.

The doctor advises a house clearing and treats the patient with a complimentary homeopathic treatment. Follow up of the patient show that he gets better within two months and is completely healed from his hay fever and associated symptoms within 10 months.

5.11. CASE: AUTHOR ANNE-MARIE: PAINFUL HIP

Method used: ACMOS Method
Symptoms: painful hip

Major corrections during balancing:
-ACMOS: essential oil - Cinnamon - acmodynamise on head: forehead. Strengthen positive memories: everything and clear negative memories.
-ACMOS: essential oil – Eucalyptus - acmodynamise on head: forehead. Strengthen certain positive memories and clear negative memories.
-ORGANIC PATHOS: Pericardium.
Acmos: essential oil – Lemon - acmodynamise on chest: Conception Vessel 15. Strengthen certain positive memories and clear negative memories
-ORGANIC PATHOS: Pericardium.
Acmos: essential oil - Basil - acmodynamise on chest: Conception Vessel 15. Strengthen positive memories and clear negative memories.
-ORGANIC PATHOS: Pericardium
Acmos: essential oil – Lemon - acmodynamise on throat: Conception Vessel 23. Strengthen certain positive memories and clear certain negative memories.
-ORGANIC PATHOS: Pericardium
Acmos: essential oil - Patchuli - acmodynamise on pelvis: Conception Vessel 4. Strengthen positive memories and clear negative memories.
-STIMULATION: All done using the Quantacmos light pen.
Marvellous vessel TCH (The Tchrong Mo): -4 Spleen, - 6 Pericardium.
Hormonal: +14 Liver, +31 Bladder, +32 Bladder, +33 Bladder, +34 Bladder ->pain in my hip is gone.
Autonomous systems: -20 Gallbladder, +10 Bladder -> Gallbladder 20 Good against lumbar conditions.
Blood: +5 Spleen.
Homework: Stomach 36: 30 seconds morning and evening; stimulate
-CHROMO REGULATION:
Colours at 2.1: no 17 (1 minute) and no 8 (5 minutes).
Colour at 17.6 Homework: no 12. Morning and night for 3 minutes.
Result: **Pain in hip is completely gone.**

CHAPTER 6: CONVERSION FORMULA

Lecher Antenna settings are for Lecher antennae with a 20,05 centimeter ruler (most commonly used ones). cm=centimeter(s); m=meter(s); s=second(s); Hz=Hertz; GHz=gigaHertz

Corresponding frequencies in Hertz can be calculated by using following formula:

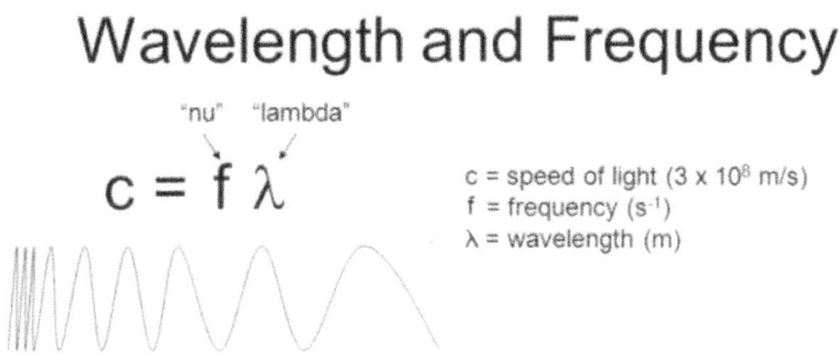

1 cm reading corresponds to 29,97 GHz; 10 cm to 2,998 GHz and 20 cm to 1,50 GHz.

The ruler of the antenna goes from 0.4cm to 18.8cm.

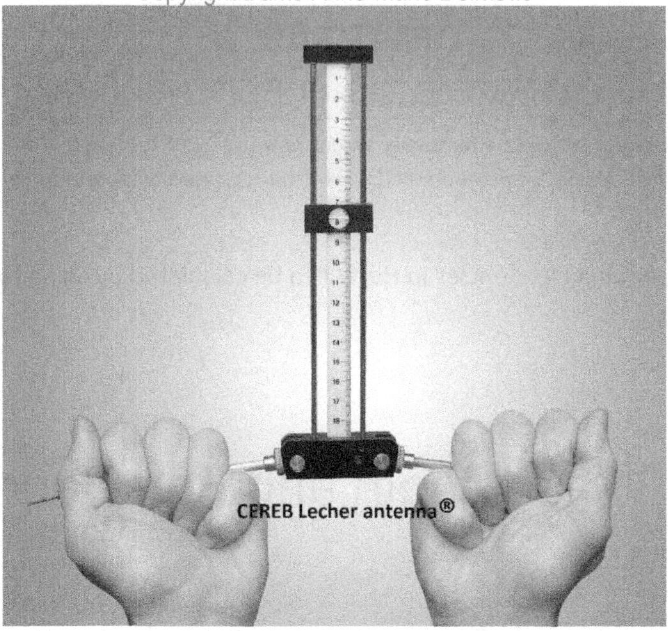

Picture: Lecher antenna CEREB® set at 7.8cm; ruler of this antenna is 20,05cm

Example: a signal at 7.8cm as in the picture above on the Lecher antenna corresponds to a frequency of:

20,05cm-7.8 cm=12,25cm distance used on the Lecher antenna=wavelength that is scanned for.

Speed of light=299792458 m/s divided by 0,1225m (12,25cm) corresponds to a frequency of 2,447 GHz.

18,8cm on Lecher antenna corresponds to a frequency of 23,98 GHz
0,4cm on Lecher antenna corresponds to a frequency of 1,526 GHz
Range for the Lecher antenna is thus from **1,526GHz to 23,98 GHz**.

GLOSSARY

-BOVIS SCALE:

The intensity of the rays or vibrations, of a place, plant or object can be measured on the Bovis Scale in Bovis units with a Lecher antenna. These are the units of vibrational quality or intensity of radiation of whatever is being measured. A healthy habitat and a healthy person should be at least 6,500 Bovis units. Values below 6,500 Bovis units indicate a qualitative energetic deficiency. Values above 6,500 Bovis units indicate a higher quality.

-COSMO-TELLURIC ENERGY:

The energy through which we connect to the earth or telluric energy is measured at 8.0 on the Lecher antenna and our cosmic connection is measured at 12.0 on the Lecher antenna. The synergy of these energies or cosmo-telluric energy is measured at 15.3 on the Lecher antenna at the height of the heart chakra.

-SPIRAL SIGNAL:

Energy lines can show a spiral signal; this means that several signals can be found coming for the same energyline; these are actually the signals of the energy spiral when looking at the energy from above.

-VERIFICATION OF AUTHENTICITY OF SIGNAL WITH MAGNETISED ROD:

Some Lecher antennae come with a magnetised rod in the left arm that can be used to verify whether a signal for a certain energy is authentic.

QUESTIONS AND ANSWERS – ANSWERS AND QUESTIONS

by Anne-Marie Delmotte

The veil to what lingers in the invisible can only be lifted
with an open mind, heart and soul

I set out on a quest for answers
I found my answer in the Lecher antenna

The answer that I found has led me to more questions
Are we to content ourselves with a wisdom that is half silent?

I, for one, will not cease my quest
Until I have found
All the answers

REFERENCES

-*Practical Guide for Dowsing with the Lecher Antenna – Elaborate Basic Training Course in Geobiology and Bio-energy* by Anne-Marie Delmotte, Edition Delmotte Vibrating Energies Résonnantes, second edition 2019.

-*E-learning course Dowsing with the Lecher Antenna Bio-energy Basic and intermediate Level* by Anne-Marie Delmotte, 2018.

-*E-learning course Dowsing with the Lecher Antenna Geobiology Basic level* by Anne-Marie Delmotte, 2018.

-*Several training courses in Brussels and Ans in Belgium* organised by CEREB, Michel Lespagnard in 2008, 2012, 2016 and 2017.

-*Training course: The Acmos Method of Bio-energetics* byRéné Nacacchian and Carol Robertson in Balloch in Scotland; Bio 1 and Bio 2, 2017.

-*Culloden 1746 – Fight for the Throne* by Chris Tabraham, Lomond Books Ltd, 2016.

-*The Megalithic Cemetery of Carrowmore Co. Sligo* by Göran Burenhult, 2001.

-https://gillighansworld.ie/

-www.knowth.com Megalithic Culture

-*Archeology in County Sligo* by Lough Arrow Research Project, Highwood Community Resource Centre, 1998.

- *Celtic Culture: A Historical Encyclopedia,* Koch, John T., ABC-CLIO, 2006.

-*Points of Cosmic Energy* by Blanche Merz; The C.W. Daniel Company Ltd, 1987.

-*Signpost to the Holy Grail? Infinite Energy with Infinite Possibilities! Dowsed with the Lecher Antenna in Carnac and Brittany France* by Anne-Marie Delmotte, Edition Delmotte, 2018.

-*The Art of Medicine – A foundation for understanding the principles of energy in medical sciences* by Thomas K. Szulc M.D., 2013.

-*The Concise Book of Acupoints* – revised edition by John R. Cross, Blue River Press, 2014.

-Earth energies and geopathic stress by Alex Stark.

Full article can be read here: http://alexstark.com/pdfs/Geopathic%20Stress.pdf

-http://www.helios3.com/electromagnetic-radiation-electrosensitivity.html

-*Les réseaux géobiologiques et leurs répercussions sur la santé* by Gilbert and Nadine Fleck and Pierre-Jean Garel ; Preface by Dr. Ernst Hartmann, Editions Jouvence, 2004.

-*Antenne de Lecher – Interprétation des taux vibratoires*, Claude François Dubuc, Édition Dubuc, 2018.

-*Magnetisme, sleutel tot het leven. Dokter met Lecher antenne*, Doctor Nestor Driscart, Drukkerij Printal, 2004.

BIOGRAPHY

Dame Anne-Marie Delmotte has an Associate's Degree in Clinical Chemistry and works as a scientist at the Belgian Government.

In 2018, she received the title of Knight in the Leopold Order from the King of Belgium for her work. The work she does for the government is completely different from her own personal energy research project.

Because she is hypersensitive to all kinds of energies and wanted to objectify what she was sensing, she searched for an instrument that is able to measure and determine these energies objectively and scientifically. In Belgium, in 2008, she was introduced to the Lecher antenna by the non-profit organisation CEREB - Belgium from whom she acquired her so long for searched instrument. She attended several training courses organised by this non-profit organisation which introduced her to geobiology and bio-energy measurements. She set off to do measurements at places that are known for their special energies like the Chartres Cathedral near Paris and several megalithic sites in Ireland. After the miraculous recovery of her mother, who was dying, after a balancing of her energies with the Lecher antenna she decided to learn more about bio-energy balancings. She found this through an additional training course in Belgium, several books and an ACMOS training course in Scotland and through communication with elaborate Lecher antenna specialists in the field.

Anne-Marie also holds following qualifications: USUI Reiki, Shamanic Reiki and Reiki Crystal Master/Teacher/Practitioner Degree as well as a Reiki Space Clearing Practitioner Degree.

OTHER BOOKS/PRODUCTS BY THE AUTHOR

In English:
-Signpost to the Holy Grail? Infinite Energy with Infinite Possibilities! Dowsed with the Lecher Antenna in Carnac and Brittany France
-Practical Guide for Dowsing with the Lecher Antenna – Elaborate Basic Training Course in Geobiology and Bio-energy

-e-learning courses: please check Facebook Page or website for special promotions or reductions
->E-learning course: Dowsing with the Lecher Antenna Bio-energy Level 1 and 2

->E-learning course: Dowsing with the Lecher Antenna Geobiology Basic Level

->E-learning course: Dowsing with the Lecher Antenna Bio-energy Harmonic Remedies Levels 1, 2, 3&4

In French:
-Signe pour le Saint Graal? Énergie infinie avec des possibilités infinies! Mesurer avec l'antenne de Lecher à Carnac et en Bretagne en France

-Guide pratique pour faire la radiesthésie avec l'antenne de Lecher - formation de base elaborée en géobiologie et bio-énergie

-cours e-learning: please check French Facebook Page or website for special promotions or reductions

Apprendre à pratiquer la radiesthésie avec l'antenne de Lecher bio-énergie et Apprendre à pratiquer la radiesthésie avec l'antenne de Lecher géobiologie

SOCIAL MEDIA

FOR FURTHER INFORMATION:

Website English: www.lecherantenna-antennedelecher.com

Website French: www.antennedelecherradiesthesie.com

Email: annemarielecherenergy@gmail.com

Facebook page (English): Lecher Antenna

Facebook page (English): Megaliths and their special energies

Facebook page (French): AntennedeLecher

Facebook page (Dutch/Flemish): Lecher antenne opleiding en boeken

Facebook group: Lecher antenna-dowsing-bioenergy-geobiology

Delmotte
Vibrating
Energies
Résonnantes ®

Made in the USA
Monee, IL
13 September 2026

62974271R00059